NATURAL PRODUCTS
INTERACTIONS
ON GENOMES

Clinical Pharmacognosy Series

Series Editors
Navindra P. Seeram
Maria Tiziana Corasaniti

Aromatherapy: Basic Mechanisms and Evidence Based Clinical Use
Giacinto Bagetta, Marco Cosentino,
and Tsukasa Sakurada

Herbal Medicines: Development and Validation of Plant-Derived Medicines for Human Health
Giacinto Bagetta, Marco Cosentino, Marie Tiziana Corasaniti,
and Shinobu Sakurada

Natural Products Interactions on Genomes
Siva G. Somasundaram

Clinical Pharmacognosy Series

NATURAL PRODUCTS INTERACTIONS ON GENOMES

Edited by

Siva G. Somasundaram, PhD

University of Houston - Victoria
Sugar Land, Texas, USA

CRC Press
Taylor & Francis Group
Boca Raton London New York

CRC Press is an imprint of the
Taylor & Francis Group, an **informa** business

First published in paperback 2024

First published 2016 by CRC Press
2385 NW Executive Center Drive, Suite 320, Boca Raton FL 33431

and by CRC Press
4 Park Square, Milton Park, Abingdon, Oxon, OX14 4RN

CRC Press is an imprint of Taylor & Francis Group, LLC

© 2016, 2024 by Taylor & Francis Group, LLC

Publisher's Note
The publisher has gone to great lengths to ensure the quality of this reprint but points out that some imperfections in the original copies may be apparent.

ISBN: 978-1-4398-7231-4 (hbk)
ISBN: 978-1-03-292615-5 (pbk)
ISBN: 978-0-429-10730-6 (ebk)

DOI: 10.1201/b19110

Visit the Taylor & Francis Web site at
http://www.taylorandfrancis.com

and the CRC Press Web site at
http://www.crcpress.com

Dedication

---

*I dedicate to my beloved sons, Dr. Aswin, MD,
and Mr. Alwin for their support.*

Contents

Series Preface

Botanical medicines are rapidly increasing in global recognition with significant public health and economic implications. For instance, in developing countries, a vast majority of indigenous populations use medicinal plants as a major form of health care. Also, in industrialized nations, including those in Europe and North America, consumers are increasingly using herbs and botanical dietary supplements as part of integrative health and complementary and alternative therapies. Moreover, the paradigm shifts occurring in modern medicine, from mono-drug to multi-drug and poly-pharmaceutical therapies, have led to renewed interest in botanical medicines and botanical drugs. This is due, in part, to the basic underpinnings of botanical medicines, according to which a complex matrix of multiple compounds within an extract elicits multiple biological actions and the activity of a major compound may be potentiated by multiple minor constituents. In short, *the whole is greater than the sum of its parts*. However, the widespread use and resurgence in the popularity of herbal medicines raises concerns about clinical efficacy, quality, safety, dosing, and the potential for herb–herb, herb–food, and herb–drug interactions. Pharmacognosy, the study of drugs from natural sources, plays a critical role in the process of ensuring the authenticity, purity, and consistency of botanicals and also for developing tools and models for determining their mechanisms and modes of action, doses, toxicity, and safety. The cross-fertilization of classical pharmacognosy with modern chemical and biological approaches, and their applications in a clinical setting, has led to the *Clinical Pharmacognosy* series, which seeks to disseminate emerging research and discuss challenges and opportunities on the aforementioned issues.

Navindra P. Seeram
University of Rhode Island

Preface

The mechanistic roles of natural products and genomes are included in the six chapters of this book. The *Dictionary of Natural Products* reported 210,000 compounds in their database. There are around 20,000 genes in human chromosomes that code for the proteins. If we want to study the impact of 210,000 compounds for each gene, then it is an endless process and a huge task. Here, the authors try to present the tip of the iceberg to understand the methods of study of the natural products and their effects on genomes. To achieve our objective, we selected major natural products and studied their interactions with randomly selected genes for each chromosome. Then we focused on genes that are involved in specific disease and presented the natural product interactions. For example, natural products and their extracts modulate mechanisms of inflammation. Inflammation is the basic pathogenesis of any chronic illness, including cancer. Most of the natural products generally inhibit acute inflammation inducing genes. For example, they fail to inhibit carcinogenesis if the inflammation becomes chronic. If we have comprehensive knowledge of the genomic interactions for all the chromosomes related to a specific natural product, it is easy to formulate a tailor-made therapy for individuals depending on their genomic profile. This approach eventually enables formulation of personalized medical therapy through clinical pharmacognosy using available natural products.

In addition, there are several reports available on the negative effects of the interactions of natural products with regular therapy. As the natural products supplements are not regulated by FDA, and available from over the counter, it is possible that the patients are using their own choice of selecting dietary supplements and mixing up their own medications without the knowledge of the side effects of drug–natural product interactions.

Here, we provide two specific studies on breast cancer and prostate cancer to address this important issue. Next, we focus on the effect of natural products on microbial growth and finally provide future perspectives on the interaction of natural products with genomes.

Throughout this volume, there are figures whose captions contain URLs to the PubMed websites, where they will find an updated literature for the respective natural products for the selected genes. In this way, this book is automatically updating the relevant bibliographical details when it is published in the PubMed and keeping abreast of the latest knowledge regarding content of interest. This will reduce the time and search for the readers and remain useful beyond the date of the publication of the book as long as the PubMed websites are active.

Siva G. Somasundaram
University of Houston-Victoria

Acknowledgments

The work of editorial assistants Ashley Weinstein and Hilary LaFoe was helpful; CRC Press was also helpful in the completion of the book. Their cooperation is much appreciated. I thank them for their daily responses to queries and collection of manuscripts and documents.

Editor

Dr. Somasundaram is the Director of Biological Sciences for Undergraduate Studies and Professor of Biology at University of Houston-Victoria, where he teaches genome sciences as part of the MS biomedical science program. He has accumulated 33 years of research experience and published 45 peer-reviewed research articles. He has been teaching biology and biochemistry for 22 years. Dr. Somasundaram received the Hari Ohm Ashram gold medals for his clinical research on diabetes and arthritis in India. He received the Enron Teaching Excellence Award 2013 and the Research Excellence Award 2010 from the University of Houston-Victoria, Sugar Land, Texas.

Dr. Somasundaram was a visiting scientist at MRC-Clinical Research Center, Harrow, United Kingdom, from 1990 to 1992. He then moved to King's College Hospital, London, as an associate clinical biochemist (1992–1997) and researched pathogenesis and prevention of intestinal inflammation induced by nonsteroidal anti-inflammatory drugs. There he has applied a novel noninvasive method to study small-intestinal permeability and absorptive capacity in HIV/AIDS, liver transplantation, cardiopulmonary bypass surgery, and inflammatory bowel disease patients.

Dr. Somasundaram moved to the University of Washington, Seattle, as a senior fellow (1997–2000) and carried out research on the transcriptional regulation and function of ileal bile acid transporter gene in human, hamster, CFTR mice, and rats by RT-PCR and immunofluorescent histochemistry. He has served as a research associate in the Lineberger Comprehensive Cancer Center, University of North Carolina, Chapel Hill, North Carolina (2000–2002) and contributed to the study of molecular signal transduction in apoptosis, mainly the inhibition of chemotherapy-induced apoptosis of human breast cancer cell lines by in vitro and in vivo xenograft animal models using dietary antioxidant supplements. He was selected as an NIH-sponsored regional fellow to conduct Inside Cancer Workshops in 2009–2010. He received USDA grants in 2006 for studying the effect of citrus limonin on the transcriptomic profiling metastasis breast cancer genome.

Contributors

Richard S. Gunasekera
Department of Biology
University of Houston-Victoria
Sugar Land, Texas

G. K. Jayaprakasha
Department of Horticultural Sciences
Texas A&M University
College Station, Texas

Bhimanagouda S. Patil
Department of Horticultural Sciences
Texas A&M University
College Station, Texas

Janet Price
Department of Cancer Biology
UT-MD Anderson Cancer Center
Houston, Texas

Chandra Somasundaram
Department of Biology
University of Houston-Victoria
Sugar Land, Texas

Siva G. Somasundaram
Department of Biology
University of Houston-Victoria
Sugar Land, Texas

and

Department of Horticultural Sciences
Texas A&M University
College Station, Texas

1 Genomics Is an Evolving Science due to the Interactions of Natural Products

Siva G. Somasundaram

CONTENTS

The genome is the entirety of an organism's *hereditary* information, and its evolution is based on the duplications of genes for an organism's survival. The environment plays a significant role in propagating genes. Mainly nutrition availability helps genomes alter and evolve to adapt in the new and emerging environment. The hereditary information is encoded within the DNA of an organism and occasionally the RNA of viruses. The genome includes both coding and non-coding DNA. The term *genome* was coined by Hans Winkler, who was a professor of botany at the University of Hamburg, Germany, in 1920. By combining the words *gene* and *chromosome*, Winkler developed the term *genome*. Until the beginning of the twentieth century, little was known about the genome. Today, thanks in large part to the Human Genome Project, genomic research has expanded, and our understanding of the evolution of the genome is growing. This new knowledge opens up the door for new fields of science to develop. Bioinformatics, for example, combines the information from sequenced genomes with computer databases to make searching for small pieces of information within complete genomic sequence almost instantaneously. By additionally examining how our food sources and natural products have developed or stayed the same over time, we will begin to understand how our foods interact with our genome. We begin our journey by exploring how the genome interact with the natural products humans consume.

1.1   GENOME AS A UNIT OF ANY BIOLOGICAL ORGANISM

Genes are basic units of heredity of all living organisms. They are areas of DNA or RNA sequences that code to specific functions. All genes are made up of four different bases: adenine, cytosine, guanine, and thymine. These are nitrogenous bases that can be divided into two smaller groups. Pyrimidines are made up of thymine and cytosine, while purines are made up of adenine and guanine. These four bases code the structures for life. Adenine bonds with thymine and guanine bonds with cytosine to form the double helix DNA structure. These codes are then stored into 23 chromosomes forming the human genome or the completed structure of the blueprint of life.

1.2   NATURAL PRODUCTS AS A FUNCTIONAL PRODUCT OF THE GENOME

Natural products are chemical products or substances formed by a living organism. They are formed naturally through chemical or biological processes. Natural products can be used for consumption and medicinal purposes. One of the most common forms of natural products is found in plants, but they can also be found in some animals and microorganisms. For centuries, natural products found in plants have been used to combat diseases. Medicinal plants have been used around the world from Chinese medicine to ancient Egypt to medicine men of Native Americans in North America. In this book, we will discuss 28 different natural products (Table 1.1) and examine whether their natural properties inhibit the diseases found on all 23 chromosomes.

The main objective of this book is to explore the interactions between natural products, disease, and modern medicine. The ultimate goal is the ability for doctors and patients to understand the pros and cons associated with the foods consumed during the disease treatment process. The understanding of the potential health risks associated with certain foods and disease will create a more informed society of people who are able to choose to consume food to better their health.

1.3   NATURAL PRODUCTS FROM ANY BIOLOGICAL ORGANISM MODULATE THE GENOME OF ANOTHER ORGANISM

Natural products are a result of gene products. In this section, we will look at a general overview of each of the 28 natural products that are being discussed in this text. These natural products will be discussed further in detail in chapters 2–5. Natural products are broadly classified into three large categories: plant derived, microorganism derived, and animal derived. We intend to focus on plant-derived natural products and their effects on humans and microorganisms. It is considered that most pathogenic micobes cause disease in humans and animals and that plant-derived natural products either prevent or inhibit the pathogenicity orginated from microbes. In that sense, we have selected a few plant-derived natural products to study the effect on both pathogenic and nonpathogenic micobes. The chemical structure of plant-derived natural products determines interactions with a genome.

TABLE 1.1
List of Natural Products with Chemical Structures

Acid ascorbic

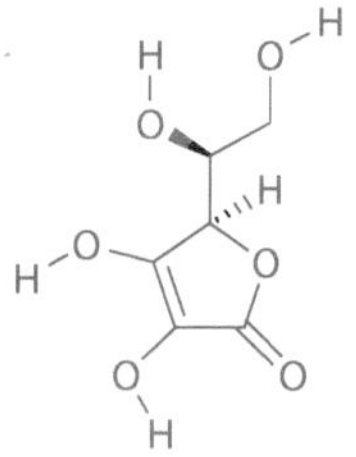

Apigenin

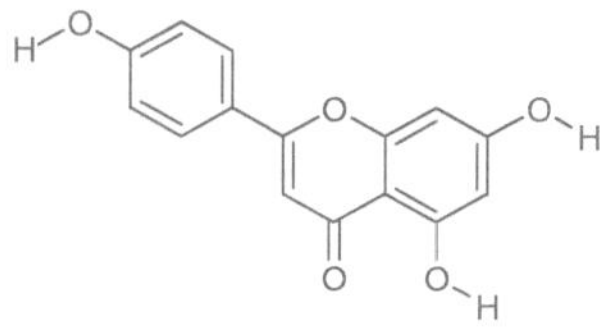

Caffeic acid

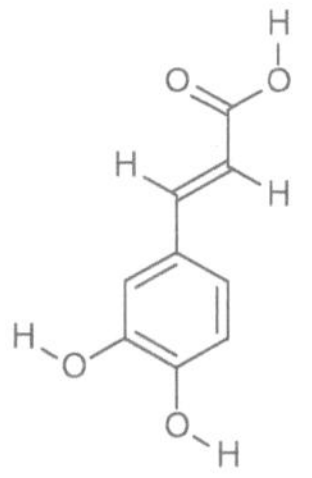

Carnosic acid

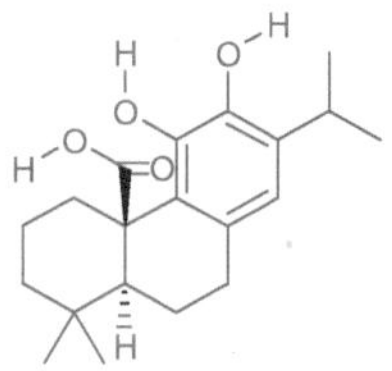

Chrysin

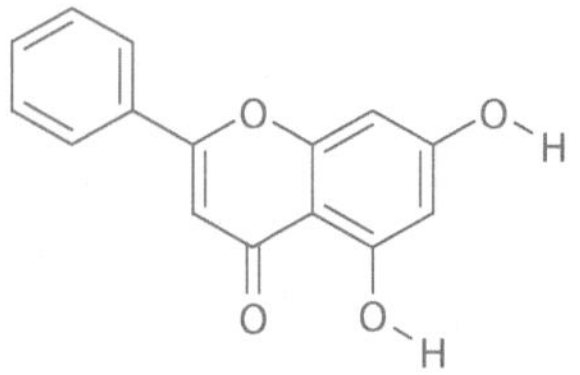

(Continued)

TABLE 1.1 (*Continued*)
List of Natural Products with Chemical Structures

Curcumin

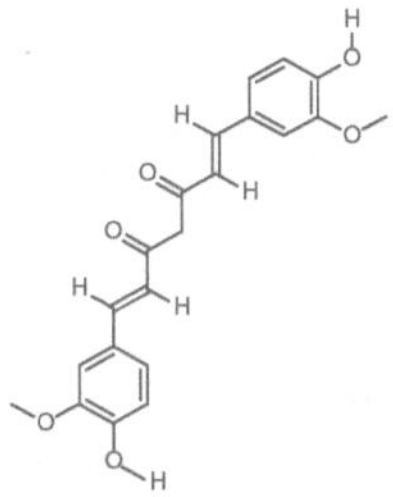

Diosmine

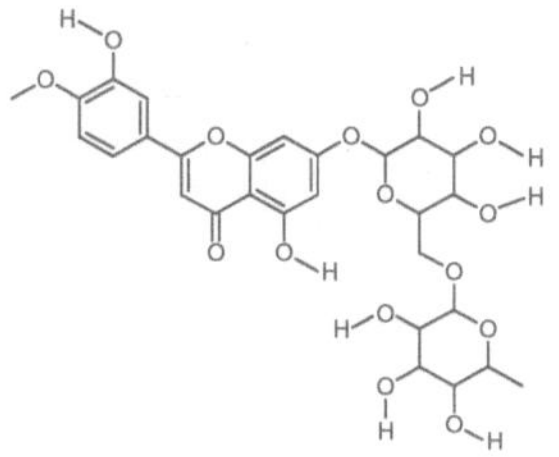

Eckol

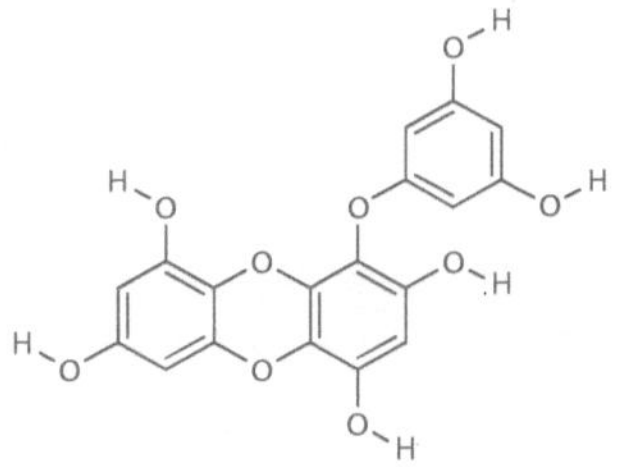

Ferulic acid

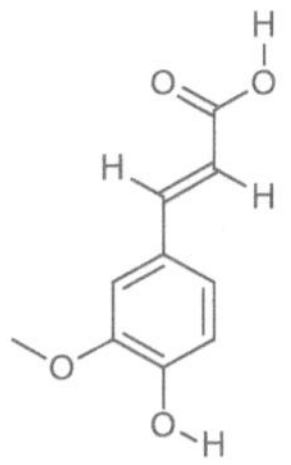

Genistein

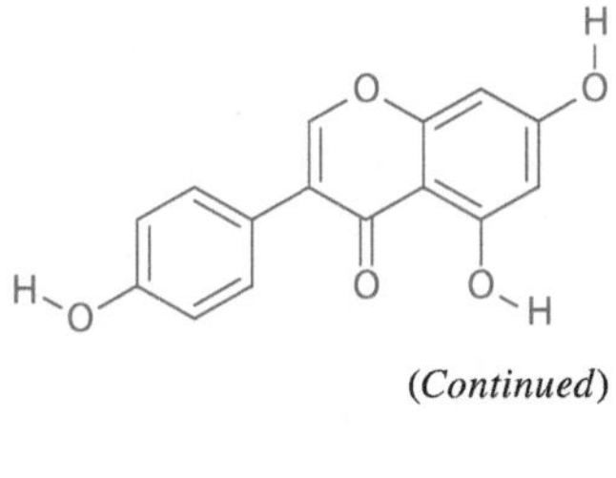

(*Continued*)

TABLE 1.1 (*Continued*)
List of Natural Products with Chemical Structures

Green tea extract
   Epigallocatechin gallate (EGCG)

Hawthorn
   Contains flavonoids such as vitexin,
   rutin, and quercetin
Hesperidin

Lycopene

Podophyllum hexandrum
   Podophyllin

Punica granatum
   Ellagic acid, the main polyphenol in
   pomegranate

(*Continued*)

TABLE 1.1 (*Continued*)
List of Natural Products with Chemical Structures

Quercetin

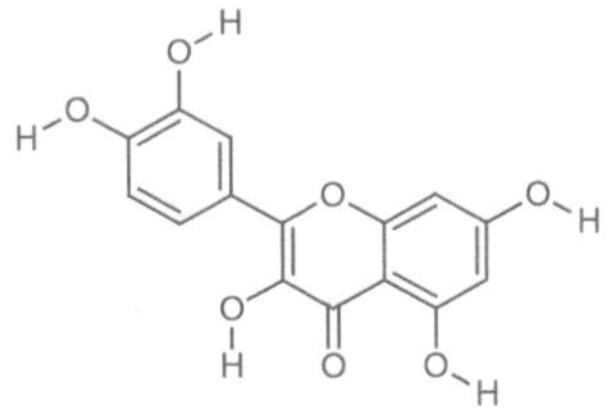

Resveratrol

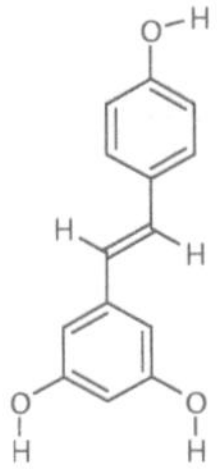

Rosmarinic acid

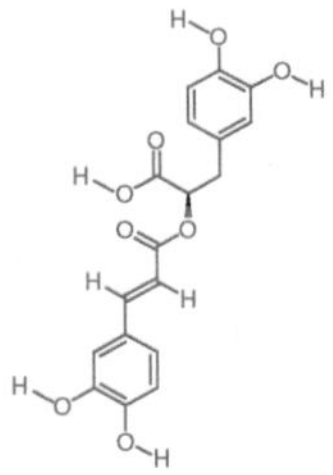

Rutin

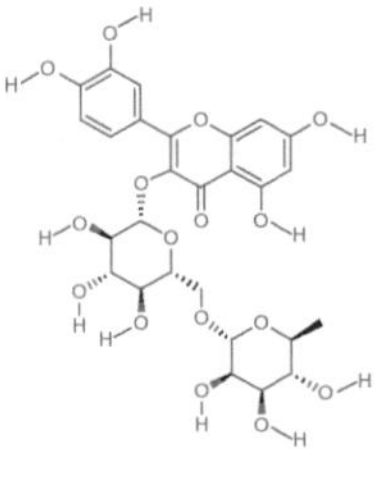

Sesamol

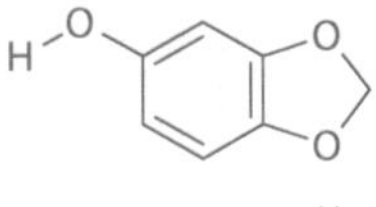

Syzygium cumini
   Eugenol

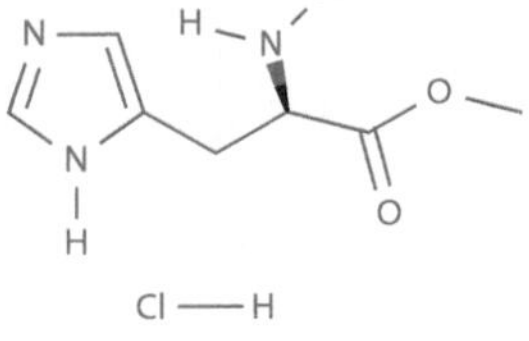

(Continued)

TABLE 1.1 (*Continued*)
List of Natural Products with Chemical Structures

Thymol

Vernonia cinerea
Vernolide-A

δ-Tocopherol

Source: The chemical structures are taken from PubChem, National Library of Medicine (https://pubchem .ncbi.nlm.nih.gov/).

1.4 CLASSIFICATION OF NATURAL PRODUCTS

Natural products can be classified on the basis of the number of carbon atoms and their arrangements to form a possible group. The following examples provide a detailed account:

1. Five-carbon atoms constitute a C_5 ring to form isoprenoids group.
2. Six-carbon atoms constitute a C_6 ring to form phenolic group.
3. Seven-carbon atoms constitute a C_6 ring with 1 C attachment arrangment to form phenolic acid group.
4. Eight-carbon atoms constitute C_6–C_2 ring to form either acetophenones or phenylacetic acid group.
5. Nine-carbon atoms constitute C_6–C_3 ring to form various groups such as cinnamic acid derivatives, phenyl propenes, coumarins, isocoumarins, and chromenes.
6. Ten-carbon atoms constitute C_6–C_4 ring to form naphthoquinones group and C_5–C_5 ring to form monoterpenes group.
7. Thirteen-carbon atoms constitute C_6–C_1–C_6 ring to form xanthone group.

8. Fourteen-carbon atoms constitute C_6–C_2–C_6 ring to form stilbenes and anthraquinones groups.

9. Fifteen-carbon atoms constitute C_6–C_3–C_6 ring to form flavonoids, isoflavonoids, and neoflavonoids groups and C_5–C_5–C_5 ring to form sesquiterpenoids group.

10. Eighteen-carbon atoms constitute a different kind of ring structure $(C_6$–$C_3)_2$ to form large lignin group.

11. Twenty-carbon atoms constitute $(C_5$–$C_5)_2$ ring to form diterpenes group.

12. Twenty-four-carbon atoms constitute C_6–C_3–C_6–C_3–C_6 ring to form a complex flavonoids group.

13. Thirty-carbon atoms constitute $(C_6$–C_3–$C_6)_2$ ring to form biflavonoids group and $(C_5$–$C_5)_3$ ring to form triterpenoids group.

14. Forty-carbon atoms constitute $(C_5$–$C_5)_4$ ring to form carotenoids group.

15. More-than-40-carbon atoms are large molecules that constitute lignins and tannins groups.

16. In addition, the alkaloids with N-containing heterocyclic ring form glycosides.

Most natural products depend on their structure to modulate cellular metabolism, directly or indirectly, to change the function of the tissues by virtue of their electron affinity to eradicate the free radicals and act as scavengers. They can also be metabolized in the liver or the site of absorption (intestine) and activated, hydroxylated by cytochrome P450 hydroxylase enzymes, or glucorodinated to excrete via the kidney or the intestine through the bile duct. During this transition, they are able to circulate in the body through blood or lymphatics. Thus, these metabolized/nonmetabolized natural products can influnce the cells of different organs and may exert their activities by either directly interacting with the respective cell membrane receptors or entering into the cells and influencing the protein or DNA or RNA synthesis machinaries. So far, the interactions of natural products with genomes are not exclusively approached. In this book, attempts are made to study the interactions of different natural products with genes, and functional proteins of individual chromosome have been cataloged using the up-to-date PubMed database.

1.5 EVOLUTION OF GENOMICS WILL GROW ALONG WITH THE REVOLUTION OF NATURAL PRODUCTS CHEMISTRY

Natural products are the basis for any diet. Without these phytochemcials our bodies could not function, but it is important to have the proper amounts of any chemical in the correct dosage. An excess of certain natural products can cause toxicity in the body. This can cause major health problems including rashes and other vital organs failure. On the other end of the spectrum, too little of a natural product can be used in dieting to lose weight. This can be healthy to a certain point. If the biochemical levels in the human body become too low, then health issues may arise. It is important to consume a complete diet with all the necessary nutrients for adequate health.

The consumption of natural products can either increase or decrease gene activity. This can work in three different stages at the cellular level. First, natural products might interact or modify the plasma membrane. Second, natural products enter the cytosol, which can either become active or reactive, or create no interaction. And finally, natural products encounter the nucleus of cells. It may bind with histones or chromosomes. This would cause changes in the function of the cells through epigenetic mechanism.

The increase or decrease of cellular bioavailability of any natural product can be used to inhibit disease or remove from the diet to avoid interactions. Many types of chemical-synthetic pharmaceuticals have shown to interact with natural products. Knowing which natural products interact with medications prescribed for disease can eliminate the interactions caused. Being aware of all interactions also gives medical professionals the ability to prescribe a specific diet tailored for individual care. In the future, this will give each patient the best-case scenario for survival as per the personal genomic data.

Some of the different effects shown by natural product interactions are the increase or decrease of gene expression. Certain foods can inhibit or enhance specific gene expression on the cellular level. The genetic makeup of an individual will determine whether or not a person is likely to be more susceptible to gene interaction. Individual natural products might also have interactions, either positive or negative, on the proteins/receptors of cellular membranes. By studying the signaling molecules, researchers are hoping to understand how the interactions might affect, activate, or modify cellular signals for effective function.

Though there have been many hundreds of natural products studied and cataloged so far, none of them have been reported on the comprehensive gene interaction profile. There are several hundreds of gene in each chromosome. Because of time constraint, we focused on only 28 natural products and tested the interaction of each compound on 10 different randomly selected genes of each chromosome. We found that some, and not all, genes are studied extensively. We eliminated the genes that have no interactions with the selected natural products. This is a small step, and in the future, it will be done for other genes that have profound effects on natural products.

In the present attempt, we explore the interactions of the listed 28 natural products with selected genes for each human chromosome. We provide the list of genes in each chromosome that are analyzed in the study. We also present the list of genetic diseases for each chromosome, as presented in National Center for Biotechnology Information (NCBI) websites. We provide unique web interactive genes of chromosomes and natural products with last-minute PubMed updated bibliographies. Then, we present the experimental findings of specific natural products interactions with genes involved in cancer and with genomes of microorganisms as a comparative genomic perspective.

2 Interactions of Natural Products with Selected Genes of Mammalian Chromosomes

Siva G. Somasundaram

CONTENTS

2.1   HUMAN CHROMOSOMES AND GENETIC DISEASES

There are several reasons for the body to become ill. The adverse reactions from diets to drugs are due to the genetic makeup of an individual. It is assumed that if we already know the effect of certain natural products from a diet's interaction with genes or its functional products such as proteins, receptors, enzymes, and neurotransmitters, then it is easy to select the advantage/disadvantage foods to consume for a healthy living. In light of personalized medicine or therapy, the knowledge of choosing the best-suitable natural products-containing diet is very important. The concept of personalized therapy depends on an individual's genome.

To enhance our understanding on this subject, it is necessary to know the existing genetic diseases that are connected to each chromosome. The following tables provide an update of genetic diseases arising from each chromosome. The selected genes of study for respective chromosomes are given following the list of genetic diseases. Next to the list of genes, a detailed natural products interacting genes chart figures with respective bibliographical details, as well as the summary of the chart is given to understand the concept. The natural products of our interest have not impacted all the genes of our selection of the respective chromosomes. Therefore, we make an effort to provide a summary of the interacting genes with only specific natural products. There are 23 chromosomes that are discussed in this chapter. If any of the natural products make an impact on the genes that cause genetic diseases, then the presented analytical results from the cell culture and animal models may help to identify a structural basis of new therapy and help personalize medicine in future.

2.2   CHROMOSOME 1

Table 2.1 summarizes 52 types of genetic diseases from chromosome 1. Table 2.2 provides the list of genes that are chosen for the natural products interactions studies. As presented in Figure 2.1, there are numerous natural products that affect genes in chromosome 1. For instance, through the activation of Na-K-Cl cotransporter (NKCC1), genistein enhances the nerve growth factor (NGF)-induced neurite outgrowth of pheochromocytoma cells (PC12). Previous studies demonstrated that soy isoflavone genistein reverses allodynia, oxidative stress, inflammation, and vascular dysfunction and enhances NGF content in diabetic mice. This suggests that genistein has the possibility of being used for therapeutic treatment of diabetes complications. Another study showed that dietary genistein also reduces the activity of thyroid peroxidase (TPO). When rats were exposed to feed with genistein, the activity of TPO rates was reduced by up to 80%. However, the remaining enzymatic activity was adequate enough to sustain thyroid homeostasis in the absence of additional perturbations.

The natural product, resveratrol, also affects NGF. A study conducted on ONS-76 medulloblastoma cells showed that resveratrol increases the expression of Zhangfei, trkA, and EGR-1, a gene typically activated by NGF–trkA signaling; this induces apoptosis and differentiation. Consequently, shortly after being treated with resveratrol, the ONS-76 cells stopped growing, indicating the potential antitumor effect. *Vernonia cinerea* also influences NGF. A bioactive compound of *V. cinerea* called 4-sulfo-benzocyclobutene induces NGF activity.

TABLE 2.1
Genetic Diseases Linked to Chromosome 1

1. 1q21.1 deletion syndrome
2. 1q21.1 duplication syndrome
3. Alzheimer's disease
4. Alzheimer's disease, type 4
5. Breast cancer
6. Brooke Greenberg Disease (Syndrome X)
7. Carnitine palmitoyltransferase II deficiency
8. Charcot–Marie–Tooth disease, types 1 and 2
9. Collagenopathy, types II and XI
10. Congenital hypothyroidism
11. Deafness, autosomal recessive deafness 36
12. Ehlers–Danlos syndrome
13. Ehlers–Danlos syndrome, kyphoscoliosis type
14. Factor V Leiden thrombophilia
15. Familial adenomatous polyposis
16. Galactosemia
17. Gaucher disease
18. Gaucher disease, type 1
19. Gaucher disease, type 2
20. Gaucher disease, type 3
21. Gaucher-like disease
22. Gelatinous drop-like corneal dystrophy
23. Glaucoma
24. Hemochromatosis
25. Hemochromatosis, type 2
26. Hepatoerythropoietic porphyria
27. Homocystinuria
28. Hutchinson–Gilford progeria syndrome
29. 3-Hydroxy-3-methylglutaryl-CoA lyase deficiency
30. Hypertrophic cardiomyopathy, autosomal dominant mutations of TNNT2; hypertrophy usually mild; restrictive phenotype may be present; may carry high risk of sudden cardiac death
31. Maple syrup urine disease
32. Medium-chain acyl-coenzyme A dehydrogenase deficiency
33. Microcephaly
34. Muckle–Wells syndrome
35. Nonsyndromic deafness
36. Nonsyndromic deafness, autosomal dominant
37. Nonsyndromic deafness, autosomal recessive
38. Oligodendroglioma
39. Parkinson's disease
40. Pheochromocytoma
41. Porphyria
42. Porphyria cutanea tarda

(Continued)

TABLE 2.1 (*Continued*)
Genetic Diseases Linked to Chromosome 1

43. Popliteal pterygium syndrome
44. Prostate cancer
45. Stickler syndrome
46. Stickler syndrome, COL11A1
47. TAR syndrome
48. Trimethylaminuria
49. Usher syndrome
50. Usher syndrome, type II
51. Van der Woude syndrome
52. Variegate porphyria

TABLE 2.2
Chromosome 1 Selected Genes

ACADM: Acyl-Coenzyme A Dehydrogenase, C-4 to C-12 Straight Chain

    This gene provides instruction for the production of coenzyme A dehydrogenase, which is important for the breakdown of medium-chain fatty acids.

COL11A1: Collagen, Type XI, Alpha 1

    This gene encodes for one of the two alpha chains of type XI collagen.

GJB3: Gap Junction Protein, Beta 3, 31kda (Connexin 31)

    This gene is part of the connexin gene family. It encodes for the Gap junction beta-3 protein that is part of gap junction.

KIF1B: Kinesin Family Member 1B

    This gene encodes for a motor protein, which transports mitochondria and synaptic vesicle precursors.

NGF: Nerve Growth Factor

    This gene encodes for a secreted protein that homodimererizes and is incorporated into a larger complex.

PARK7: Parkinson's Disease (Autosomal Recessive, Early Onset) 7

    This gene acts as a positive regulator of androgen receptor-dependent transcription.

GLC1A: Gene for Glaucoma

    This gene encodes for myocilin.

HPC1: Gene for Prostate Cancer

    This gene is a genetic locus.

IRF6: Gene for Connective Tissue Formation

    The gene is part of the interferon regulatory transcription factor family.

ASPM: A Brain-Size Determinant

    This gene encodes for the abnormal spindle-like microcephaly-associated protein.

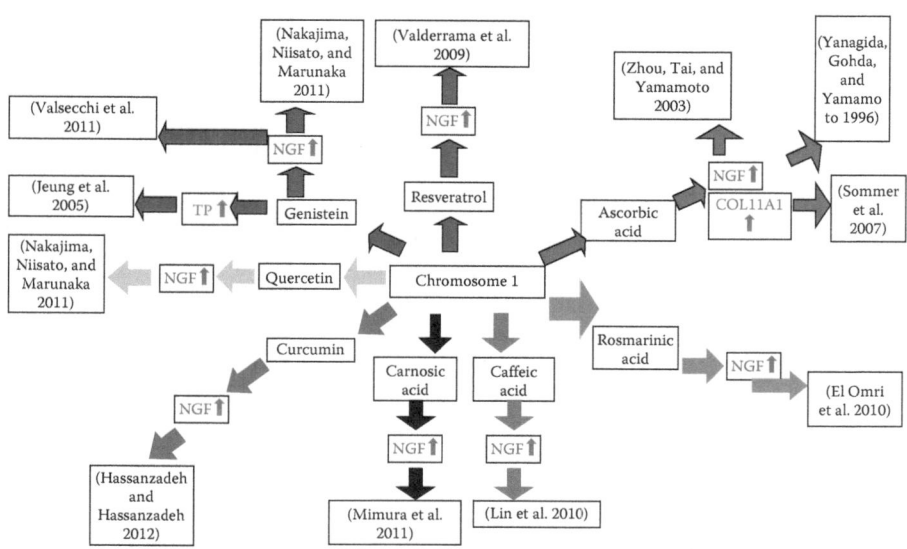

FIGURE 2.1 (See color insert.) Natural products interacting with the selected genes of chromosome 1. COL11A1 [COL11A1 and ascorbic acid]: http://www.ncbi.nlm.nih.gov/pubmed/?term=COL11A1+AND+Ascorbic+Acid; NGF [NGF and ascorbic acid]: http://www.ncbi.nlm.nih.gov/pubmed/?term=NGF+AND+Ascorbic+Acid; NGF [NGF and caffeic acid]: http://www.ncbi.nlm.nih.gov/pubmed/?term=NGF+AND+Caffeic+Acid; NGF [NGF and carnosic acid]: http://www.ncbi.nlm.nih.gov/pubmed/?term=NGF+AND+Carnosic+Acid; NGF [NGF and curcumin]: http://www.ncbi.nlm.nih.gov/pubmed/?term=NGF+AND+Curcumin; NGF [NGF and genistein]: http://www.ncbi.nlm.nih.gov/pubmed/?term=NGF+AND+Genistein; NGF [NGF and quercetin]: http://www.ncbi.nlm.nih.gov/pubmed/?term=NGF+AND+Quercetin; NGF [NGF and resveratrol]: http://www.ncbi.nlm.nih.gov/pubmed/?term=NGF+AND+Resveratrol; NGF [NGF and rosmarinic acid]: http://www.ncbi.nlm.nih.gov/pubmed/?term=NGF+AND+Rosmarinic+Acid; and TP [TP and genistein]: http://www.ncbi.nlm.nih.gov/pubmed/?term=TP+AND+Genistein. (Data from El Omri, A. et al., *J Ethnopharmacol*, 131, 451–8, 2010; Hassanzadeh, P., and A. Hassanzadeh, *Neurochem Res*, 37, 1112–20, 2012; Jeung, H. C. et al., *Biochem Pharmacol*, 70, 13–21, 2005; Lin, W. L. et al., *Chem Biol Interact*, 188, 607–15, 2010; Mimura, J. et al., *J Biochem*, 150, 209–17, 2011; Nakajima, K. et al., *Biomed Res*, 32, 351–6, 2011; Nakajima, K. et al., *Cell Physiol Biochem*, 28, 147–56, 2011; Sommer, F. et al., *Tissue Eng*, 13, 1281–9, 2007; Valderrama, X. et al., *J Neurooncol*, 91, 7–17, 2009; Valsecchi, A. E. et al., *Eur J Pharmacol*, 650, 694–702, 2011; Yanagida, M. et al., *Life Sci*, 59, 2075–81, 1996; Zhou, X. et al., *Biol Pharm Bull*, 26, 341–6, 2003.)

Ascorbic acid also has an effect on NGF. In the presence of ascorbic acid, NGF has a synergetic effect on inducing neurtie outgrowth in PC12 cells. Another study found that the addition of 2-*O*-alpha-D-glucopyranosyl-L-ascorbic acid (AA-2) in synergy with NGF causes the stimulation of a primary antigen-specific antibody response and thus enhances the immune system. An additional gene, which is affected by ascorbic acid, is COL11A1. Ascorbic acid also enhances the expression COL11A1 in hyaolcytes in the eye. It modulates proliferation and collagen accumulation of hyalocyte as well as affects the mRNA expression of the cells.

Rosmarinic acid affects the NGF gene. It displays neurotrophic effects in PC12 cells via cell differentiation induction and cholinergic activities. Caffeic acid influences NGF in such a way that it has antitumor progression potential. The authors showed that caffeic acid phenethyl ester (CAPE) inhibits the metastasis of C6 glioma cells in nude mice. It stimulated the expression of NGF and p75 neurotrophin receptor, which are involved in neural cell differentiation. Carnosic acid generates NGF production in human T98G glioblastoma cells as well as normal human astrocytes. Curcumin's positive effect on NGF suggests that it has potential as a therapeutic medicine. This group showed that a four-week treatment with curcumin resulted in a sustained elevation of brain NGF and endocannabinoids. Quercetin also affects NGF. It stimulates the NGF-induced neurite outgrowth; this is done through an increase in Cl⁻ incorporation into the intracellular space by activating NKCC1 in PC12 cell. Therefore, the selected natural products mainly have an important impact on a particular gene, NGF, and exhibit their molecular interactions in chromosome 1, and the interactions of these selected natural products with other genes of chromosome 1 have not been discussed in this book.

2.3   CHROMOSOME 2

Table 2.3 summarizes 29 types of genetic diseases from chromosome 2. Table 2.4 provides the list of genes that are chosen for the natural products interactions studies. As presented in Figure 2.2, the effects of five natural products on chromosome 2 genes were examined. To start with, the natural product, curcumin, has a significant effect on the MSH2, a tumor suppressor gene. Curcumin activates the accumulation of DNA double-strand breaks (DSB) and the induction of a checkpoint response via mismatch repair (MMR)-dependent mechanism. In MMR-comprised cells, curcumin-induced DSB is significantly blunted; consequently, cells fail to experience cell arrest, enter mitosis, and die through mitotic catastrophe. Thus, curcumin could have therapeutic value, particularly in treatment of tumors with compromised MMR function. Another natural product, which has an impact on the MSH2 gene, is quercetin. This study showed that quercetin considerably downregulates the potentially oncogenic mitogen-activated protein kinase (MAPK) pathway. Moreover, quercetin enhances the expression of tumor suppressor genes, including MSH2. Accordingly, tumor-protective mechanisms are correlated with a shift in energy production pathways, signifying a decreased cytoplasmic glycolysis and increased mitochondrial fatty acid degradation.

Apigenin also affects a gene in chromosome 2, and it affects the TPO gene. The authors showed that apigenin found in fonio millet demonstrates strong anti-TPO activities. This effect results in a major reduction of hormonogenic capacity of the enzyme in osteoblasts and human bone marrow mesenchymal stem cells. Evidently, this natural product has beneficial interactions with genes in chromosome 2. Genistein also affects the TPO gene according to a study conducted on rats that showed that dietary genistein reduces the activity of TPO. In this study, rats were fed with genistein, and as a result, the activity of TPO rates was significantly reduced. The remaining enzymatic activity was adequate enough to sustain thyroid homeostasis. In addition, ascorbic acid has a significant impact on COL3A1 gene because

TABLE 2.3
Genetic Diseases Linked to Chromosome 2

1. 2p15-16.1 microdeletion syndrome
2. Autism
3. Alport syndrome
4. Alström syndrome
5. Amyotrophic lateral sclerosis
6. Amyotrophic lateral sclerosis, type 2
7. Congenital hypothyroidism
8. Dementia with Lewy bodies
9. Ehlers–Danlos syndrome
10. Ehlers–Danlos syndrome, classical type
11. Ehlers–Danlos syndrome, vascular type
12. Fibrodysplasia ossificans progressiva
13. Harlequin type ichthyosis
14. Hemochromatosis
15. Hemochromatosis, type 4
16. Hereditary nonpolyposis colorectal cancer
17. Infantile-onset ascending hereditary spastic paralysis
18. Juvenile primary lateral sclerosis
19. Long-chain 3-hydroxyacyl-coenzyme A dehydrogenase deficiency
20. Maturity onset diabetes of the young, type 6
21. Mitochondrial trifunctional protein deficiency
22. Nonsyndromic deafness
23. Nonsyndromic deafness, autosomal recessive
24. Primary hyperoxaluria
25. Primary pulmonary hypertension
26. Sitosterolemia (knockout of either ABCG5 or ABCG8)
27. Sensenbrenner syndrome
28. Synesthesia
29. Waardenburg syndrome

TABLE 2.4
Chromosome 2 Selected Genes

NCL: Nucleolin

This gene is involved in the synthesis and maturation of ribosomes.

COL3A1: Collagen Type III, Alpha 1 (Ehlers–Danlos Syndrome Type IV, Autosomal Dominant)

This gene encodes for a fibrillar collagen, which is found in connective tissues.

CTLA4: Cytotoxic T-Lymphocyte Antigen 4

This gene encodes for a protein that transmits an inhibitory signal to T cells.

TPO: Thyroid Peroxidase

This gene encodes for a membrane-bound glycoprotein that acts as an enzyme and has an important role in thyroid gland function.

(Continued)

TABLE 2.4 (*Continued*)
Chromosome 2 Selected Genes

TBR1: T-Box, Brain, 1
  This gene encodes transcription factors involved in the regulation of developmental processes.
BMPR2: Bone Morphogenetic Protein Receptor, Type II (Serine/Threonine Kinase)
  This gene encodes for a member of the bone morphogenetic protein receptor family of
  transmembrane serine and threonine kinases.
MSH2: Muts Homolog 2, Colon Cancer, Nonpolyposis, Type 1
  This gene is a protein-coding gene.
SLC40A1: Solute Carrier Family 40 (Iron-Regulated Transporter), Member 1
  This gene encodes a cell membrane protein, which may probably be involved in iron export from the
  duodenal epithelial cells.
PAX3: Paired Box Gene 3 (Waardenburg Syndrome 1)
  This gene is associated with ear, eye, and facial development.
NR4A2: Nuclear Receptor Subfamily 4, Group A, Member 2
  This gene encodes for the nuclear receptor-related 1 protein.

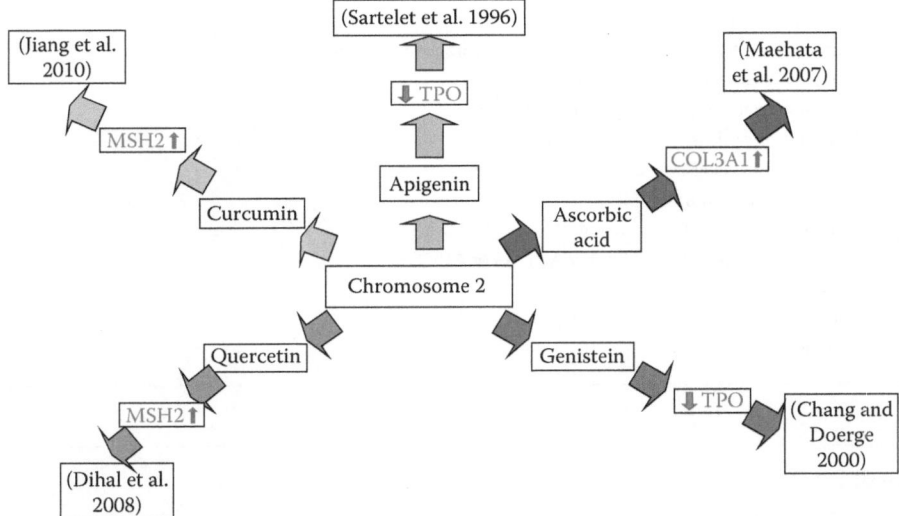

FIGURE 2.2   (See color insert.) Natural products interacting with the selected genes of
chromosome 2. COL3A1 [COL3A1 and ascorbic acid]: http://www.ncbi.nlm.nih.gov/pubmed/?
term=COL3A1+AND+Ascorbic+acid; MSH2 [MSH2 and curcumin]: http://www.ncbi.nlm.
nih.gov/pubmed/?term=MSH2+AND+curcumin; MSH2 [MSH2 and quercetin]: http://www.
ncbi.nlm.nih.gov/pubmed/?term=MSH2+AND+Quercetin; TPO [TPO and apigenin]: http://
www.ncbi.nlm.nih.gov/pubmed/?term=TPO+AND+Apigenin; and TPO [TPO and genis-
tein]: http://www.ncbi.nlm.nih.gov/pubmed/?term=TPO+AND+Genistein. (Data from Chang,
H. C., and D. R. Doerge, *Toxicol Appl Pharmacol*, 168, 244–52, 2000; Dihal, A. A. et al., *Proteomics*, 8, 45–61, 2008; Jiang, Z. et al., *Mol Cancer Ther*, 9, 558–68, 2010; Maehata,
Y. et al., *Matrix Biol*, 26, 371–81, 2007; Sartelet, H. et al., *Nutrition*, 12, 100–6, 1996.)

type III collagen is critical for growth acceleration of human osteoblastic cells via ascorbic acid 2-phosphate. Ascorbic acid 2-phosphate stimulates cell growth and the expression of mRNA for type III collagen in human osteoblast-like MG-63 cells, and the interactions of these selected natural products with other genes of chromosome 2 have not been discussed in this book.

2.4 CHROMOSOME 3

Table 2.5 summarizes 70 types of genetic diseases linked to chromosome 3. Table 2.6 provides the list of genes that are chosen for the natural products interactions studies. As presented in Figure 2.3, the effects of two different natural products on genes in chromosome 3 have been studied. Green tea regulates phospholipase D-1 (PLD1). According to a study, it is revealed that a major component of green tea, epigallocatechin-3 gallate (EGCG), regulates PLD activity in human astroglioma cells. Curcumin affects a different gene called GATA2 that drives the development and progression of prostate cancer. This study showed that curcumin suppresses the function of these pioneer

---

TABLE 2.5

Genetic Diseases Linked to Chromosome 3

1. 3-Methylcrotonyl-CoA carboxylase deficiency
2. 3q29 microdeletion syndrome
3. Alkaptonuria
4. Arrhythmogenic right ventricular dysplasia
5. Atransferrinemia
6. Autism
7. Biotinidase deficiency
8. Blepharophimosis, epicanthus inversus, and ptosis, type 1
9. Breast/colon/lung/pancreatic cancer
10. Brugada syndrome
11. Castillo fever
12. Carnitine-acylcarnitine translocase deficiency
13. Cataracts
14. Cerebral cavernous malformation
15. Charcot–Marie–Tooth disease, type 2
16. Charcot–Marie–Tooth disease
17. Chromosome 3q duplication syndrome
18. Coproporphyria
19. Dandy Walker syndrome
20. Deafness
21. Diabetes
22. DA receptor
23. Dystrophic epidermolysis bullosa
24. Endplate acetlycholinesterase deficiency
25. Essential tremors
26. Glaucoma, primary open angle

(Continued)

TABLE 2.5 (*Continued*)
Genetic Diseases Linked to Chromosome 3

27. Glycogen storage disease
28. Hailey–Hailey disease
29. Harderoporphyrinuria
30. Heart block, progressive/nonprogressive
31. Hereditary coproporphyria
32. Hereditary nonpolyposis colorectal cancer
33. HIV infection, susceptibility/resistance to
34. Hypobetalipoproteinemia, familial
35. Hypothermia
36. Leukoencephalopathy with vanishing white matter
37. Long QT syndrome
38. Lymphomas
39. Malignant hyperthermia susceptibility
40. Metaphyseal chondrodysplasia, Murk Jansen type
41. Microcoria
42. Moebius syndrome
43. Moyamoya disease
44. Mucopolysaccharidosis
45. Muir–Torre family cancer syndrome
46. Myotonic dystrophy, type 2
47. Myotonic dystrophy
48. Neuropathy, hereditary motor, and sensory, Okinawa type
49. Night blindness
50. Nonsyndromic deafness, autosomal recessive
51. Nonsyndromic deafness
52. Ovarian cancer
53. Porphyria
54. Propionic acidemia
55. Protein S deficiency
56. Pseudo-Zellweger syndrome
57. Retinitis pigmentosa
58. Romano-Ward syndrome
59. Sensenbrenner syndrome
60. Septo-optic dysplasia
61. Short stature
62. Spinocerebellar ataxia
63. Sucrose intolerance
64. T cell leukemia translocation altered gene
65. Usher syndrome, type III
66. Usher syndrome (Finland)
67. Usher syndrome
68. von Hippel–Lindau syndrome
69. Waardenburg syndrome
70. Xeroderma pigmentosum, complementation group c

TABLE 2.6
Chromosome 3 Selected Genes

PLD1: Phospholipase D1

This gene encodes for a phosphatidylcholine-specific phospholipase that catalyzes the hydrolysis of phosphatidylcholine to yield phosphatidic acid and choline.

MCM2: Minichromosome Maintenance Complex Component 2

This gene encodes for one of the highly conserved mini-chromosome maintenance proteins, which are involved in the start-up of eukaryotic genome replication.

FANCD2: Fanconi Anemia, Complementation Group D2

This gene encodes for the protein of complementation group D2.

ACPP: Acid Phosphatase, Prostate

This gene encodes for an enzyme, which catalyzes the conversion of orthophosphoric monoester to alcohol and orthophosphate.

GATA2: GATA-Binding Protein 2

The gene encodes for a protein that is part of the GATA family of zinc-finger transcription factors.

RAB5A: RAB5A, Member RAS Oncogene Family

This gene is needed for the fusion of plasma membranes and early endosomes.

MST1R: Macrophage-Stimulating 1 Receptor (C-Met-Related Tyrosine Kinase)

This gene encodes for a cell surface receptor for macrophage-stimulating protein with tyrosine kinase activity.

PEX5L: Peroxisomal Biogenesis Factor 5-Like

This gene is an important part of hyperpolarization-activated cyclic nucleotide-gated channels, which regulate the cell-surface expression and cyclic nucleotide dependence.

TNFSF10: Tumor Necrosis Factor Superfamily, Member 10

This gene encodes for a cytokine, which is part of the tumor necrosis factor ligand family.

PIK3CA: G Elongation Factor, Mitochondrial 1

This gene is oncogenic.

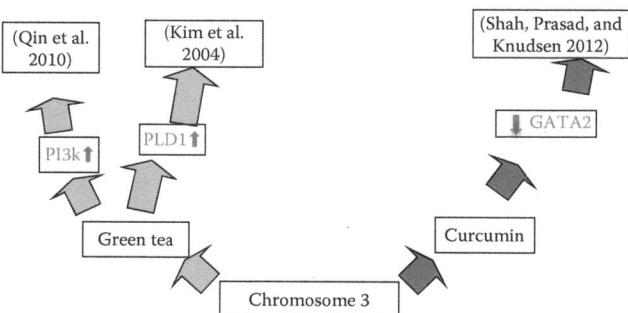

FIGURE 2.3 **(See color insert.)** Natural products interacting with the selected genes of chromosome 3. GATA2 [GATA2 and curcumin]: http://www.ncbi.nlm.nih.gov/pubmed/?term= GATA2+AND+Curcumin; PI3k [PI3k and green tea]: http://www.ncbi.nlm.nih.gov/pubmed/? term=Qin+et+al.+2010+green+tea; and PLD1 [PLD1 and green tea]: http://www.ncbi. nlm.nih.gov/pubmed/?term=PLD1+AND+Green+Tea. (Data from Kim, S. Y. et al., *Eur J Biochem*, 271, 3470–80, 2004; Qin, B. et al., *Mol Nutr Food Res*, 54, S14–23, 2010; Shah, S. et al., *Cancer Res*, 72, 1248–59, 2012.)

factors, leading to a reduction in tumor growth, and suggests that curcumin affects the GATA2 gene in such a way that it suppresses tumor progression. These studies demonstrate that both green tea and curcumin have a significant impact on the genes in chromosome 3, and the interactions of these selected natural products with other genes of chromosome 3 have not been discussed in this book.

2.5   CHROMOSOME 4

Table 2.7 summarizes 27 types of genetic diseases from chromosome 4. Table 2.8 provides the list of genes that are chosen for the natural products interactions studies. As presented in Figure 2.4, four different natural products demonstrated significant effects on the genes in chromosome 4. For instance, green tea causes an increase on the expression of the IP-10 gene. The authors found that mice treated with a green tea EGCG showed an increase in the antifibrotic protein IP-10. This resulted

---

TABLE 2.7
Genetic Diseases Linked to Chromosome 4

1. Achondroplasia
2. Autosomal dominant polycystic kidney disease (PKD-2)
3. Bladder cancer
4. Crouzonodermoskeletal syndrome
5. Chronic lymphocytic leukemia
6. Ellis-van Creveld syndrome
7. Facioscapulohumeral muscular dystrophy
8. Fibrodysplasia ossificans progessiva FOP
9. Hemophilia C
10. Huntington's disease
11. Hemolytic uremic syndrome
12. Hirschprung's disease
13. Hypochondroplasia
14. Methylmalonic acidemia
15. Muenke syndrome
16. Nonsyndromic deafness
17. Nonsyndromic deafness, autosomal dominant
18. Ondine's curse
19. Parkinson's disease
20. Polycystic kidney disease
21. Romano–Ward syndrome
22. SADDAN
23. Tetrahydrobiopterin deficiency
24. Thanatophoric dysplasia
25. Thanatophoric dysplasia, type 1
26. Thanatophoric dysplasia, type 2
27. Wolfram syndrome

TABLE 2.8
Chromosome 4 Selected Genes

IP-10: Interferon Gamma-Induced Protein 10
    This gene encodes for a chemokine of the CXC subfamily and ligand for the CXCR3 receptor.
CBR4: Carbonyl Reductase 4
    This gene is a mitochondrial NADPH-dependent reductase for o- and p-quinones.
FGB: Fibrinogen Beta Chain
    This gene encodes for the beta component of fibrinogen, which is a blood-borne glycoprotein comprising three pairs of nonidentical polypeptide chains.
GAB1: Grb2-Associated-Binding Protein 1
    This gene encodes for a member of the IRS1-like multi-substrate docking protein family.
MIP-2: Macrophage Inflammatory Protein 2
    The gene encodes for a cytokine of the CXC chemokine family.
NPNT: Nephronectin
    This gene is associated with acute kidney tubular necrosis and renal agenesis.
PTTG2: Pituitary Tumor-Transforming 2
    This gene is associated with prostatitis and pituitary tumors.
RNF4: Ring Finger Protein 4
    This gene encodes for a protein that has a ring finger motif and acts as a transcription regulator.
SLBP: Stem Loop-Binding Protein
    This gene encodes for a protein, which binds to the stem-loop structure in replication-dependent histone mRNAs.
SFRP2: Secreted Frizzled-Related Protein 2
    This gene encodes for a protein that is part of the SFRP family containing a cysteine-rich domain homologous to the putative WNT-binding site of Frizzled proteins.

in antifibrotic effects. Consequently, the study suggests that EGCG has significant potential in ameliorating the development of obliterative airway disease. Lycopene affects the MIP-2 gene. An in vitro study showed that tomato lycopene extract (TLE) blocked lipopolysaccharides (LPS)-induced transcriptional activity and MIP-2 mRNA accumulation in IEC-18 cells.

Furthermore, caffeic acid inhibits the production of interferon-gamma-inducible protein (IP-10). CAPE inhibits tumor necrosis factor (TNF)- and LPS-induced IP-10 production in a dose-dependent fashion, independently of p38 MAPK, HO-1, and Nrf2 signaling pathways in mouse intestinal epithelial cells. The authors concluded from these findings that CAPE shows anti-inflammatory potential in intestinal epithelium. Genistein increases sFRP2 expression through the inhibition of β-catenin-mediated WNT signaling and by demethylating its silenced promoter in the human colon cancer cell line. As a result, the genistein treatment reduced human cancer cell line DLD-1 cell viability and increased apoptosis. The impact of all these natural products on chromosome 4 genes is noteworthy and shows potential for treatment of various diseases, and the interactions of these selected natural products with other genes of chromosome 4 have not been discussed in this chapter.

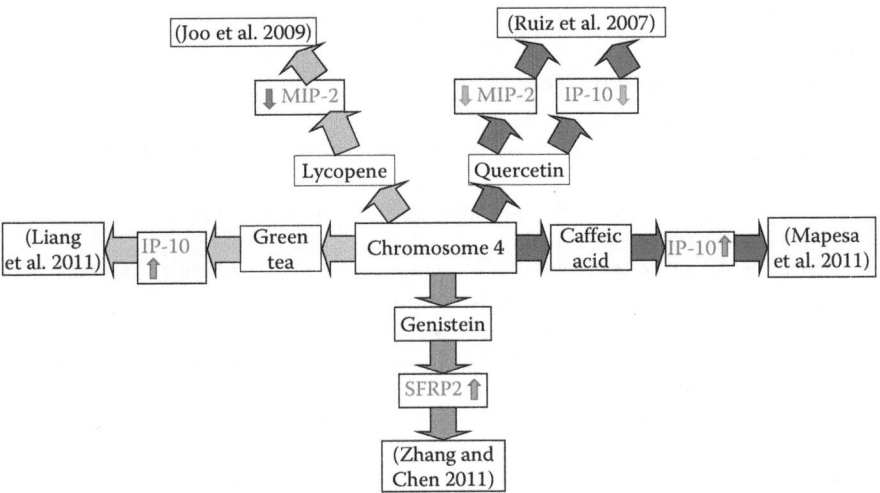

FIGURE 2.4 **(See color insert.)** Natural products interacting with the selected genes of chromosome 4. IP-10 [IP-10 and caffeic acid]: http://www.ncbi.nlm.nih.gov/pubmed/?term=IP-10+AND++Caffeic+Acid; IP-10 [IP-10 and quercetin]: http://www.ncbi.nlm.nih.gov/pubmed/?term=IP-10+AND+Quercetin; IP-10 [IP-10 and green tea]: http://www.ncbi.nlm.nih.gov/pubmed/?term=IP-10+AND+Green+Tea; MIP-2 [MIP-2 and lycopene]: http://www.ncbi.nlm.nih.gov/pubmed/?term=MIP-2+AND++Lycopene; MIP-2 [MIP-2 and quercetin]: http://www.ncbi.nlm.nih.gov/pubmed/?term=MIP-2+AND+Quercetin; and SFRP2 [SFRP2 and genistein]: http://www.ncbi.nlm.nih.gov/pubmed/?term=SFRP2+AND++Genistein. (Data from Joo, Y. E. et al., *PLoS One*, 4, e4562, 2009; Liang, O. D. et al., *Exp Lung Res*, 37, 435–44, 2011; Mapesa, J. O. et al., *Mol Nutr Food Res*, 55, 1850–61, 2011; Ruiz, P. A. et al., *J Nutr*, 137, 1208–15, 2007; Zhang, Y., and H. Chen, *Exp Biol Med [Maywood]*, 236, 714–22, 2011.)

2.6 CHROMOSOME 5

Table 2.9 summarizes 29 types of genetic diseases from chromosome 5. Table 2.10 provides the list of genes that are chosen for the natural products interactions studies. As presented in Figure 2.5, research has been done in which two genes from chromosome 5 have been affected by natural products. Those two genes are the APC gene and the EGR-1 gene.

TABLE 2.9

Genetic Diseases Linked to Chromosome 5

1. Achondrogenesis, type 1B
2. Atelosteogenesis, type II
3. Cockayne syndrome
4. Cornelia de Lange syndrome
5. Corneal dystrophy of Bowman layer, type I
6. Corneal dystrophy of Bowman layer, type II
7. Cri du Chat
8. Diastrophic dysplasia

(Continued)

TABLE 2.9 (*Continued*)

Genetic Diseases Linked to Chromosome 5

9. Ehlers–Danlos syndrome
10. Ehlers–Danlos syndrome, dermatosparaxis type
11. Familial adenomatous polyposis
12. Granular corneal dystrophy, type I
13. Granular corneal dystrophy, type II
14. GM2-gangliosidosis, AB variant
15. Homocystinuria
16. 3-Methylcrotonyl-CoA carboxylase deficiency
17. Netherton syndrome
18. Nicotine dependency
19. Parkinson's disease
20. Primary carnitine deficiency
21. Recessive multiple epiphyseal dysplasia
22. Sandhoff disease
23. Spinal muscular atrophy
24. Sotos syndrome
25. Survival motor neuron spinal muscular atrophy
26. Treacher Collins syndrome
27. Tricho-hepato-enteric syndrome
28. Usher syndrome
29. Usher syndrome, type II

TABLE 2.10

Chromosome 5 Selected Genes

APC: Adenomatous polyposis coli.

This gene encodes a tumor suppressor protein that acts as an antagonist of the Wnt signaling pathway. It is also involved in other processes including cell migration and adhesion, transcriptional activation, and apoptosis. Defects in this gene cause familial adenomatous polyposis (FAP), an autosomal dominant premalignant disease that usually progresses to malignancy. Disease-associated mutations tend to be clustered in a small region designated the mutation cluster region (MCR) and result in a truncated protein product.

EGR-1: Early growth response 1

The protein encoded by this gene belongs to the EGR family of C2H2-type zinc-finger proteins. It is a nuclear protein and functions as a transcriptional regulator. The products of target genes it activates are required for differentitation and mitogenesis. Studies suggest this is a cancer suppressor gene.

PROP1: Paired-Like Homeobox 1

This gene encodes for a paired-like homeodomain transcription factor in the developing pituitary gland.

ADAMTS2: Metallopeptidase with Thrombospondin Type 1, Motif 2

This gene encodes for a protein that is part of the ADAMTS protein family.

ALDH7A1: Aldehyde Dehydrogenase 7 Family, Member A1

This gene encodes for a protein that is a member of subfamily 7 in the aldehyde dehydrogenase gene family.

(Continued)

TABLE 2.10 (*Continued*)
Chromosome 5 Selected Genes

CTNND2: Catenin (Cadherin-Associated Protein), Delta 2

This gene encodes for a protein that is active in the nervous system, where it helps in cell adhesion and is important in cell movement.

ERCC8: Excision Repair Cross-Complementation Group 8

This gene encodes for the Cockayne syndrome A protein, which is involved in repairing damaged DNA.

FAM134: Family with Sequence Similarity 134, Member B

This gene is associated with lung cancer.

GLRA1: Glycine Receptor, Alpha 1

This gene encodes for a receptor that mediates postsynaptic inhibition in the central nervous system.

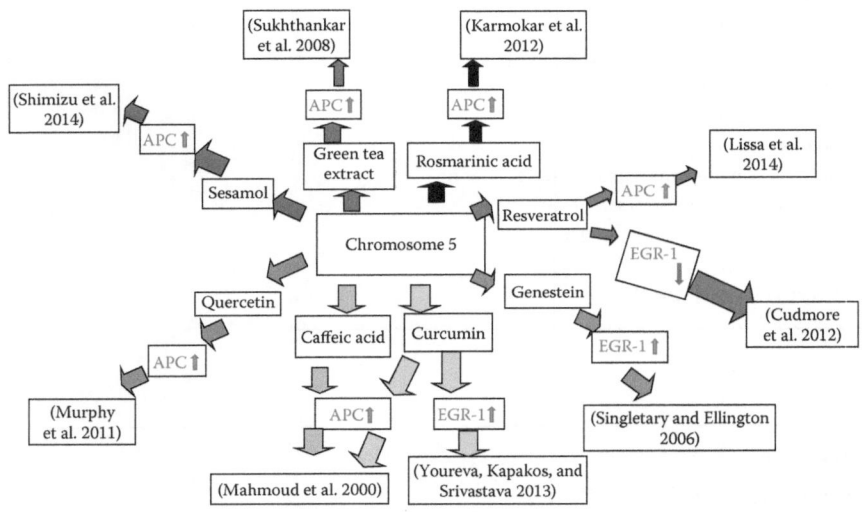

FIGURE 2.5 (**See color insert.**) Natural products interacting with the selected genes of chromosome 5. APC [APC and caffeic acid]: http://www.ncbi.nlm.nih.gov/pubmed/?term=APC +AND+Caffeic+Acid; APC [APC and green tea]: http://www.ncbi.nlm.nih.gov/pubmed/ ?term=APC+AND+Green+tea; APC [APC and quercetin]: http://www.ncbi.nlm.nih.gov/ pubmed/?term=APC+AND+Quercetin; APC [APC and resveratrol]: http://www.ncbi.nlm.nih. gov/pubmed/?term=APC+AND+Resveratrol; APC [APC and rosmarinic acid]: http://www.ncbi. nlm.nih.gov/pubmed/?term=APC+AND+Rosmarinic+Acid; APC [APC and sesamol]: http://www. ncbi.nlm.nih.gov/pubmed/?term=APC+AND+Sesamol; EGR-1 [EGR-1 and curcumin]: http:// www.ncbi.nlm.nih.gov/pubmed/?term=EGR-1+AND+curcumin; EGR-1 [EGR-1 and genestein]: http://www.ncbi.nlm.nih.gov/pubmed/?term=EGR-1+AND+Genestein; and EGR-1 [EGR-1 and resveratrol]: http://www.ncbi.nlm.nih.gov/pubmed/?term=EGR-1+AND+Resveratrol. (Data from Cudmore, M. J. et al., *Am J Obstet Gynecol*, 206, 253.e10–5, 2012; Karmokar, A. et al., *Mol Nutr Food Res*, 56, 775–83, 2012; Lissa, D. et al., *Proc Natl Acad Sci U S A*, 111, 3020–5, 2014; Mahmoud, N. N. et al., *Carcinogenesis*, 21, 921–7; 2000; Murphy, E. A. et al., 2011, *Nutr Cancer*, 63, 421–6, 2000; Shimizu, S. et al., *J Clin Biochem Nutr*, 54, 95–101, 2014; Singletary, K., and A. Ellington, *Anticancer Res*, 26, 1039–48, 2006; Sukhthankar, M. et al., *Gastroenterology*, 134, 1972–80, 2008; Youreva, V. et al., *Can J Physiol Pharmacol*, 91, 241–7, 2013.)

Several natural products affect the APC gene. The APC gene is associated with colon cancer. CAPE was able to decrease tumor formation in mice. Curcumin showed to have induced similar tumor inhibition. The mice treated with curcumin and CAPE led to an increased associated enterocyte apoptosis and proliferation and showed antitumor effect. This concluded that curcumin and CAPE are potential anti-inflammatory agents suppressing the APC-associated intestinal carcinogenesis.

Quercetin reduced polyp number and size distribution in mice with the APC gene. Next, sesamol had an effect on the APC gene and demonstrated the decrease in number of polyps in the middle part of the small intestine of APC-deficient mice. This suggests that sesamol may have an anticarcinogenetic property.

In addition to these natural products, green tea extract is one that affects the APC gene. EGCG makes up for about half of the catechin content in green tea extract. EGCG was found to inhibit the tumor formation in APC mice. Moreover, rosmarinic acid also showed to have an effect on the APC gene. Rosmarinic acid decreased the frequency of large adenomas in APC mice. This suggests that rosmarinic acid may be able to slow down adenoma development. Next, resveratrol was able to reduce the formation of tetraploid or higher-order polyploidy cells in APC mice. Oral treatment with resveratrol reduced the accumulation of tetraploid intestinal epithelial cells in the APC mouse model of colon cancer. In addition, resveratrol affects the EGR-1 gene on chromosome 5 by preventing the upregulation of EGR-1. Also, EGR-1 is affected by genestein. These are the natural products' effects on chromosome 5 genes. Many more research needs to be done to continue the search of understanding the genes on this chromosome.

2.7 CHROMOSOME 6

Table 2.11 summarizes 23 types of genetic diseases from chromosome 6. Table 2.12 provides the list of genes that are chosen for the natural products interactions studies. As presented in Figure 2.6a and b, there are a plethora of natural products, which

---

TABLE 2.11

Genetic Diseases Linked to Chromosome 6

1. Ankylosing spondylitis, HLA-B
2. Collagenopathy, types II and XI
3. Coeliac disease, HLA-DQA1 and DQB1
4. Ehlers–Danlos syndrome, classical, hypermobility, and Tenascin-X types
5. Hashimoto's thyroiditis
6. Hemochromatosis
7. Hemochromatosis, type 1
8. 21-Hydroxylase deficiency
9. Maple syrup urine disease
10. Methylmalonic acidemia
11. Autosomal nonsyndromic deafness
12. Otospondylomegaepiphyseal dysplasia
13. Parkinson's disease
14. Polycystic kidney disease

(Continued)

TABLE 2.11 (*Continued*)
Genetic Diseases Linked to Chromosome 6

15. Porphyria
16. Porphyria cutanea tarda
17. Rheumatoid arthritis, HLA-DR
18. Stickler syndrome, COL11A2
19. Systemic lupus erythematosus
20. Diabetes mellitus type 1, HLA-DR, DQA1, and DQB1
21. X-linked sideroblastic anemia
22. Epilepsy
23. Guillain Barre syndrome

TABLE 2.12
Chromosome 6 Selected Genes

ARG1: Arginase 1
   This gene is associated with hyperargininemia and recessive dystrophic epidermolysis bullosa.
CUL7: Cullin 7
   This gene encodes for a protein that is a component of an E3 ubiquitin-protein ligase complex.
HLA-B: Major Histocompatibility Complex, Class I, B
   This gene is part of the HLA class I heavy chain paralogs.
PARK2: Parkinson Protein 2 (The function of the gene is unknown.)
   It encodes for a protein that is a component of a multi-protein E3 ubiquitin ligase complex that
   mediates the targeting of substrate proteins for proteasomal degradation.
ZFP57: Zinc-Finger Protein 57
   This gene encodes for a zinc-finger protein that contains a KRAB domain.
RUNX2: Runt-Related Transcription Factor 2
   This gene is part of the RUNX family of transcription factors. It encodes for a nuclear protein with a
   Runt DNA-binding domain.
VEGFA: Vascular Endothelial Growth Factor A
   This gene is part of the PDGF/VEGF growth factor family. It encodes for a protein that is often
   found as a disulfide-linked homodimer.
HFE: Hemochromatosis
   This gene encodes for a protein that is similar to the MHC class I-type protein.
EYA4: Eyes Absent Homolog 4
   This gene encodes for a member of the eyes absent family proteins.
NEU1: Sialidase 1 (Lysosomal Sialidase)
   This gene encodes for a lysosomal enzyme that cleaves terminal sialic acid residues from substrates.

affect RUNX2 and VEGFA genes in chromosome 6. One of these natural products is genistein. Genistein enhances the mRNA level of RUNX2. Genistein also inhibits the growth of colon cancer cells through apoptosis induction and cell cycle arrest at G2/M phase. The antitumor mechanisms of genistein, include apoptosis, and cell cycle arrest through the downregulation of VEGFA. Ascorbic acid also affects both

RUNX2 and VEGFA genes. This study found that the combined effect of ascorbic acid and inorganic phosphate (Pi) on vascular smooth muscle cells resulted in a four-fold increase of Cuff $\alpha 1$/RUNX2 mRNA expression and displayed that ascorbic acid reduces VEGFA activity and tumor growth.

Resveratrol also influences RUNX2 and VEGFA genes. A study revealed that resveratrol, considerably, enhanced osteogenesis by increasing the expression of RUNX2. As a result, resveratrol promoted osteogenic differentiation of mesenchymal stem cells. In addition, another study showed that resveratrol has an antitumor effect of VEGFA gene on human renal cancer (786-0) cells. The authors found that resveratrol suppressed the expression of VEGF genes and thus significantly inhibited the proliferation of 786-0 cells. Curcumin also interacts with both RUNX2 and VEGFA genes. Curcumin prevented an increase in Cbfa1/RUNX2 expression in primary vascular smooth muscle cells, and the authors suggest that curcumin could be beneficial in the management of vascular calcification. Furthermore, it is demonstrated that curcumin greatly reduced hVEGF-A$_{165}$ overexpression to normal in Clara cells of the lungs of transgenic mice. Accordingly, curcumin was able to suppress the formation of tumors.

Caffeic acid interacts with both VEGF and RUNX2 genes. A study showed that induced VEGF expression was completely abrogated by CAPE. Caffeic acid reverses the decrease of nuclear phosphorylation of RUNX2. Quercetin also impacts VEGF and RUNX2. In one study, quercetin suppressed the key transcription factors for

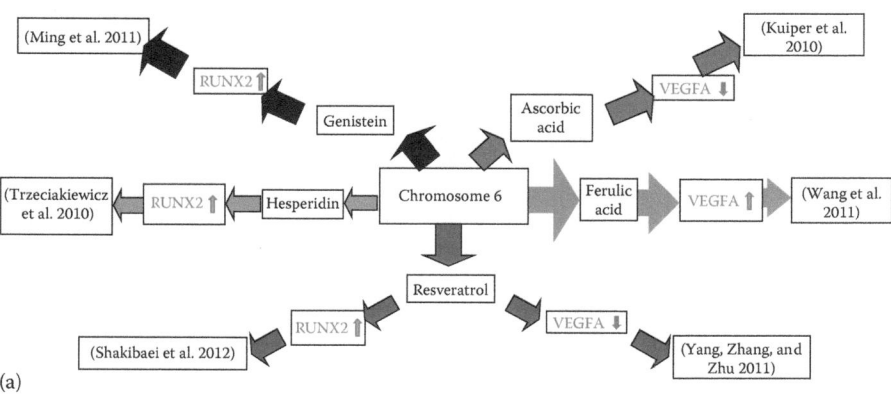

(a)

FIGURE 2.6 (See color insert.) Natural products interacting with the selected genes of chromosome 6. (a) RUNX2 [RUNX2 and genistein]: http://www.ncbi.nlm.nih.gov/pubmed/?term=RUNX2+AND+Genistein; RUNX2 [RUNX2 and hesperidin]: http://www.ncbi.nlm.nih.gov/pubmed/?term=RUNX2+AND+Hesperidin; RUNX2 [RUNX2 and resveratrol]: http://www.ncbi.nlm.nih.gov/pubmed/?term=RUNX2+AND+Resveratrol; VEGFA [VEGFA and ascorbic acid]: http://www.ncbi.nlm.nih.gov/pubmed/?term=VEGFA+AND+Ascorbic+Acid; VEGFA [VEGFA and ferulic acid]: http://www.ncbi.nlm.nih.gov/pubmed/?term=VEGFA+AND+Ferulic+Acid; and VEGFA [VEGFA and resveratrol]: http://www.ncbi.nlm.nih.gov/pubmed/?term=VEGFA+AND+Resveratrol. (Data from Kuiper, C. et al., *Cancer Res*, 70, 5749–58, 2010; Ming, L. et al., *Zhongguo Zhong Yao Za Zhi*, 36, 2240–5, 2011; Shakibaei, M. et al., *PLoS One*, 7, e35712, 2012; Trzeciakiewicz, A. et al., *J Agric Food Chem*, 58, 668–75, 2010; Wang, J. et al., *J Ethnopharmacol*, 137, 992–7, 2011; Yang, R. et al., *Mol Med Rep*, 4, 981–3, 2011.) *(Continued)*

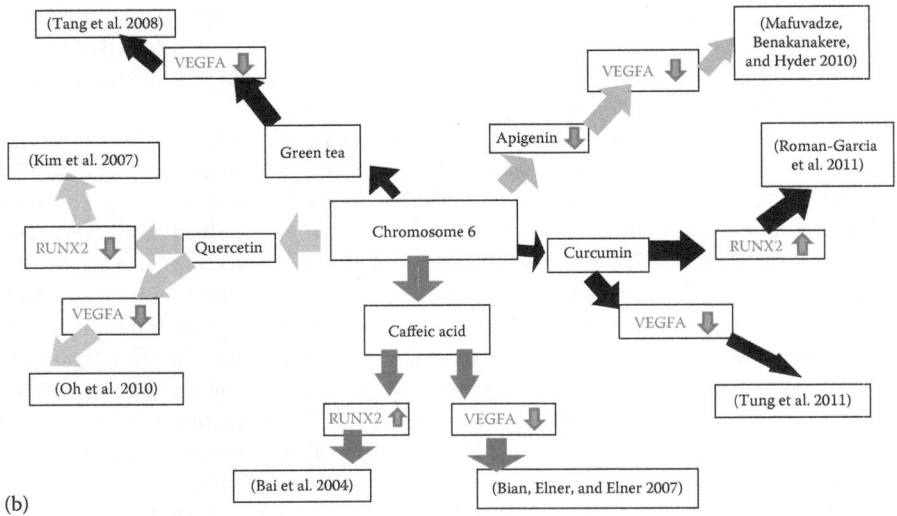

(b)

FIGURE 2.6 (Continued) Natural products interacting with the selected genes of chromosome 6. (b) RUNX2 [RUNX2 and caffeic acid]: http://www.ncbi.nlm.nih.gov/pubmed/?term =RUNX2+AND+Caffeic++Acid; RUNX2 [RUNX2 and curcumin]: http://www.ncbi.nlm.nih. gov/pubmed/?term=RUNX2+AND+curcumin; RUNX2 [RUNX2 and quercetin]: http://www. ncbi.nlm.nih.gov/pubmed/?term=RUNX2+AND+Quercetin; VEGFA [VEGFA and apigenin]: http://www.ncbi.nlm.nih.gov/pubmed/?term=VEGFA+AND+Apigenin; VEGFA [VEGFA and caffeic acid]: http://www.ncbi.nlm.nih.gov/pubmed/?term=VEGFA+AND+Caffeic++Acid; VEGFA [VEGFA and curcumin]: http://www.ncbi.nlm.nih.gov/pubmed/?term=VEGFA+AND +Curcumin; VEGFA [VEGFA and green tea]: http://www.ncbi.nlm.nih.gov/pubmed/?term= VEGFA+AND+Green+Tea; and VEGFA [VEGFA and quercetin]: http://www.ncbi.nlm.nih. gov/pubmed/?term=VEGFA+AND+Quercetin. (Data from Bai, X. C. et al., *Biochem Biophys Res Commun*, 314, 197–207, 2004; Bian, Z. M. et al., *Exp Eye Res*, 84, 812–22, 2007; Kim, D. S. et al., *J Cell Biochem*, 101, 790–800, 2007; Mafuvadze, B. et al., *Menopause*, 17, 1055– 63, 2010; Oh, S. J. et al., *Food Chem Toxicol*, 48, 3227–34, 2010; Roman-Garcia, P. et al., *J Nephrol*, 24, 669–72, 2011; Tang, X. D. et al., *Zhonghua Yi Xue Za Zhi*, 88, 2872–7, 2008; Tung, Y. T. et al., *Mol Nutr Food Res*, 55, 1036–43, 2011.)

VEGF gene transcription, and it inhibited the enhanced VEGF secretion in TAMR-MCF7 cells. These results show that quercetin might have therapeutic potential for the treatment of TAM-resistant breast cancer. Quercetin also upregulated Cbfa1/ RUNX2 mRNA expression and further confirmed that this natural product may be beneficial in chemotherapy-resistant breast cancers.

Ferulic acid interacts with the VEGFA gene. It enhanced cyclin VEGF mRNA expression in ECV304 cells and promoted the proliferation of ECV304 cells. Green tea also influences the VEGFA gene through its major component—EGCG. EGCG can significantly inhibit the VEGF protein and decrease the VEGF protein expression in human cervical carcinoma cells. Apigenin also has an effect on the VEGFA gene through blocking progestin-dependent induction of VEGF mRNA and protein. In addition, it broadly inhibits the ability of progestins to alter the expression of other components of the angiogenesis pathway, such as VEGF receptors, in human breast cancer cells.

Hesperidin interacts with the RUNX2 gene as well. Hesperetin-7-O-glucuronide (Hp7G) significantly induced mRNA expression of ALP, RUNX2, and Osterix after 48 hours of exposure. The results of this study suggest that Hp7G could regulate osteoblast differentiation through RUNX2 and Osterix stimulation and may be involved in the regulation of osteoblast/osteoclast communication. Clearly, there is a great deal of natural products that affect the VEGF and RUNX2 genes in chromosome 6, and the interactions of these selected natural products with other genes of chromosome 6 have not been discussed in this chapter.

2.8 CHROMOSOME 7

Table 2.13 summarizes 34 types of genetic diseases from chromosome 7. Table 2.14 provides the list of genes that are chosen for the natural products interactions studies. As presented in Figure 2.7, there are many natural products that affect the genes in chromosome 7. One of these natural products is hesperidin that influences the MyoD gene. Hesperidin augmented the nuclear localization and myogenin promoter

---

TABLE 2.13

Genetic Diseases Linked to Chromosome 7

1. Argininosuccinic aciduria
2. Cerebral cavernous malformation
3. Charcot–Marie–Tooth disease
4. Charcot–Marie–Tooth disease, type 2
5. Citrullinemia
6. Congenital bilateral absence of vas deferens
7. Cystic fibrosis
8. Distal spinal muscular atrophy, type V
9. Ehlers–Danlos syndrome
10. Ehlers–Danlos syndrome, arthrochalasia type
11. Ehlers–Danlos syndrome, classical type
12. Hemochromatosis
13. Hemochromatosis, type 3
14. Hereditary nonpolyposis colorectal cancer
15. Lissencephaly
16. Maple syrup urine disease
17. Maturity onset diabetes of the young, type 2
18. Mucopolysaccharidosis, type VII or Sly syndrome
19. Myelodysplastic syndrome
20. Nonsyndromic deafness
21. Nonsyndromic deafness, autosomal dominant
22. Nonsyndromic deafness, autosomal recessive
23. Osteogenesis imperfecta
24. Osteogenesis imperfecta, type I
25. Osteogenesis imperfecta, type II
26. Osteogenesis imperfecta, type III

(Continued)

TABLE 2.13 (*Continued*)

Genetic Diseases Linked to Chromosome 7

27. Osteogenesis imperfecta, type IV
28. p47-phox-deficient chronic granulomatous disease
29. Pendred syndrome
30. Romano–Ward syndrome
31. Shwachman–Diamond syndrome
32. Schizophrenia
33. Tritanopia or tritanomaly color blindness
34. Williams syndrome

TABLE 2.14

Chromosome 7 Selected Genes

AASS: Aminoadipate-Semialdehyde Synthase
 This gene encodes for an enzyme found in most tissues with the highest amounts in the liver.
ASL: Argininosuccinate Lyase
 This gene encodes for an enzyme that participates in the urea cycle.
eNOS: Nitric Oxide Synthase
 This gene is associated with susceptibility to coronary spasm
POR: P450 (cytochrome) oxidoreductase
 This gene encodes for an enzyme that is needed for the normal functioning of more than 50
 enzymes in the cytochrome P450 family.
ABCB4: ATP-Binding Cassette, Subfamily B (MDR/TAP), Member 4
 This gene encodes for a member of the superfamily of ATP-binding cassette transporters.
IL6: Interleukin 6
 This gene encodes for a cytokine, which functions in the inflammation and maturation of B cells.
CYP3A4: Cytochrome P450, Family 3, Subfamily A, Polypeptide 4
 This gene encodes for a protein that is part of the cytochrome P450 superfamily of enzymes.
EPO: Erythropoietin
 This gene encodes for a protein that is part of the EPO/TPO family and encodes a secreted,
 glycosylated cytokine composed of four-alpha helical bundles.
MET: Met Proto-Oncogene
 This gene is associated with papillary renal carcinoma.
LEP: Leptin
 This gene encodes for a protein that is secreted by white adipocytes and is important for the regulation
 of body weight.

binding of MyoD, thus enhancing MyoD-induced myogenin gene transcription. Moreover, hesperidin amplified myogenin and muscle creatine kinase gene expression during myogenic differentiation from C3H10T1/2 mesenchymal stem cells in a MyoD-dependent manner. This accelerated in vivo muscle regeneration is induced by muscle injury. Green tea influences the IL6 gene. In one study, rats were fed a high-fructose diet with green tea polyphenol (GTP) supplementation. As a result, the

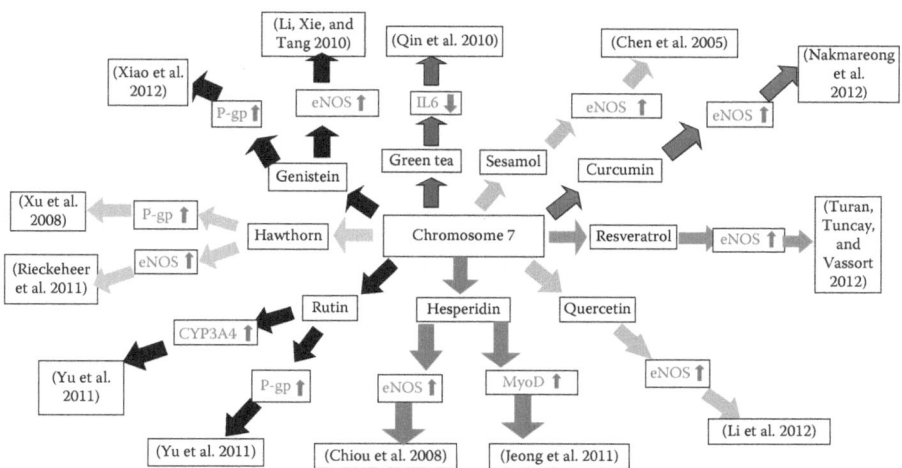

FIGURE 2.7 (See color insert.) Natural products interacting with the selected genes of chromosome 7. CYP3A4 [CYP3A4 and rutin]: http://www.ncbi.nlm.nih.gov/pubmed/?term=CYP3A4+AND+Rutin; eNOS [eNOS and curcumin]: http://www.ncbi.nlm.nih.gov/pubmed/?term=eNOS+AND+Curcumin; eNOS [eNOS and genistein]: http://www.ncbi.nlm.nih.gov/pubmed/?term=eNOS+AND+Genistein; eNOS [eNOS and hawthorn]: http://www.ncbi.nlm.nih.gov/pubmed/?term=eNOS+AND+Hawthorn; eNOS [eNOS and quercetin]: http://www.ncbi.nlm.nih.gov/pubmed/?term=eNOS+AND+Quercetin; eNOS [eNOS and resveratrol]: http://www.ncbi.nlm.nih.gov/pubmed/?term=eNOS+AND+Resveratrol; eNOS [eNOS and sesamol]: http://www.ncbi.nlm.nih.gov/pubmed/?term=eNOS+AND+Sesamol; IL6 [IL6 and green tea]: http://www.ncbi.nlm.nih.gov/pubmed/?term=IL6+AND+green+tea; MyoD [MyoD and hesperidin]: http://www.ncbi.nlm.nih.gov/pubmed/?term=MyoD+AND+Hesperidin; eNOS [eNOS and hesperidin]: http://www.ncbi.nlm.nih.gov/pubmed/?term= eNOS+AND+Hesperidin; P-gp [P-gp and genistein]: http://www.ncbi.nlm.nih.gov/pubmed/?term=P-gp+AND+Genistein; P-gp [P-gp and hawthorn]: http://www.ncbi.nlm.nih.gov/pubmed/?term=P-gp+AND+Hawthorn; and P-gp [P-gp and rutin]: http://www.ncbi.nlm.nih.gov/pubmed/?term=P-gp+AND+Rutin. (Data from Chen, P. R. et al., *Lipids*, 40, 955–61, 2005; Chiou, C. S. et al., *Clin Exp Pharmacol Physiol*, 35, 938–43, 2008; Jeong, H. et al., *Br J Pharmacol*, 163, 598–608, 2011; Li, J. et al., *Pharmacology*, 86, 240–8, 2010; Li, P. G. et al., *Pharmacology*, 89, 220–8, 2012; Nakmareong, S. et al., *Hypertens Res*, 35, 418–25, 2012; Qin, B. et al., *Mol Nutr Food Res*, 54, S14–23, 2010; Rieckeheer, E. et al., *Phytomedicine*, 19, 20–4, 2011; Turan, B. et al., *J Bioenerg Biomembr*, 44, 281–96, 2012; Xiao, C. Q. et al., *Xenobiotica*, 42, 173–8, 2012; Xu, Y. A., *Drug Dev Ind Pharm*, 34, 164–70, 2008; Yu, C. P. et al., *J Agric Food Chem*, 59, 4644–8, 2011.)

GTP decreased inflammatory factors including IL6 mRNA levels, ultimately reducing the inflammation of cardiac muscle in rats.

Rutin interacts with both P-gp and CYP3A4 genes. In vitro studies revealed that rutin induced the functions of P-gp and CYP3A4. By activating P-gp and CYP3A4, rutin decreased the bioavailability of cyclosporine. Consequently, it is recommended that transplant patients treated with Neoral (cyclosporine) should avoid consuming rutin to reduce the risk of allograft rejection. Genistein affects both P-gp and eNOS genes. Genistein induced P-gp activity in healthy volunteers. Also, genistein greatly enhanced phosphorylated eNOS and left the total eNOS, and the eNOS dimer/monomer ratio relatively unchanged. Moreover, genistein reduced the binding of eNOS with caveolin 3 and, simultaneously, elevated its binding with calmodulin and heat shock protein 90.

A plethora of other natural products affect the eNOS gene. For instance, it was discovered that treatment with the natural product, resveratrol, greatly reduced the blood glucose level in streptosotozin (STZ)-treated type 1 diabetic animals via both insulin-dependent and insulin-independent pathways. The reason is that resveratrol activates some of the similar intracellular insulin signaling components in myocardium such as eNOS. Quercetin also interacts with eNOS. Quercetin rapidly phosphorylates eNOS at Ser1179 through an Akt-independent, cAMP/PKA-mediated pathway to increase the production of nitric oxide (NO) and to promote vasodilation. A major metabolite of curcumin, tetrahydrocurcumin, increases aortic eNOS expression.

Next, sesamol induces the transcription of eNOS. Hesperidin influences the eNOS gene as well. The authors found that hesperidin treatment of human umbilical vein endothelial cells (HUVEC) enhanced NO eNOS activity and the phosphorylation of eNOS and Akt. Clearly, these natural products have a substantial influence on the genes in chromosome 7, and most of them impact the eNOS gene. The interactions of these selected natural products with other genes of chromosome 7 have not been discussed in this chapter.

2.9   CHROMOSOME 8

Table 2.15 summarizes 19 types of genetic diseases from chromosome 8. Table 2.16 provides the list of genes that are chosen for the natural products interactions studies. As presented in Figure 2.8, several natural products have interacted with the genes in chromosome 8. For instance, apigenin affects the StAR gene. Apigenin has the potential to improve StAR protein expression and steroidogenic sensitivity of aging Leydig cells. Genistein also impacts the StAR gene. A study revealed that

---

TABLE 2.15
Genetic Diseases Linked to Chromosome 8

1. 8p23.1 duplication syndrome
2. Burkitt's lymphoma
3. Charcot–Marie–Tooth disease
4. Charcot–Marie–Tooth disease, type 2
5. Charcot–Marie–Tooth disease, type 4
6. Cleft lip and palate
7. Cohen syndrome
8. Congenital hypothyroidism
9. Lipoprotein lipase deficiency, familial
10. Primary microcephaly
11. Hereditary multiple exostoses
12. Pfeiffer syndrome
13. Rothmund–Thomson syndrome, or poikiloderma congenitale
14. Schizophrenia, associated with 8p21-22 locus
15. Waardenburg syndrome
16. Werner syndrome

(Continued)

TABLE 2.15 (*Continued*)

Genetic Diseases Linked to Chromosome 8

17. Pingelapese blindness
18. Langer–Giedion syndrome
19. Roberts syndrome

TABLE 2.16

Chromosome 8 Selected Genes

LPL: Lipoprotein Lipase
  This gene encodes for lipoprotein lipase, which is expressed in heart, muscle, and adipose tissue.
IDO1: Indoleamine 2,3-dioxygenase 1
  This gene encodes for a heme enzyme that catalyzes the first and rate-limiting step in tryptophan
  catabolism to *N*-formyl-kynurenine.
StAR: Steroidogenic Acute Regulatory Protein
  This gene encodes for a protein that is important for the acute regulation of steroid hormone
  synthesis by enhancing the conversion of cholesterol into pregnenolone.
NRG1: Neuregulin 1
  This gene encodes for a protein that is identified as a 44-kD glycoprotein, which interacts with the
  NEU/ERBB2 receptor tyrosine kinase to increase its phosphorylation on tyrosine residues.
SGCZ: Sarcoglycan, Zeta
  This gene encodes for a protein that is part of the sarcoglycan complex, a group of six proteins.
SNTG1: Syntrophin, Gamma 1
  This gene encodes for a protein that is part of the syntrophin family.
CLVS1: Clavesin 1
  This gene is associated with Duane retraction syndrome and hepatocellular carcinoma.
STAU2: Staufen Double-Stranded RNA-Binding Protein 2
  This gene is associated with Becker muscular dystrophy and muscular dystrophy.
DLC1: Deleted in Liver Cancer 1
  This gene encodes for a GTPase-activating protein that is part of the rhoGAP family proteins and
  that is important for the regulation of small GTP-binding proteins.
NKAIN3: Na^+/K^+ Transporting ATPase Interacting 3
  This gene is part of the mammalian proteins that are similar to *Drosophila* NKAIN.

genistein significantly reduced StAR promoter activity in transiently transfected MA-10 cells. Quercetin increased StAR mRNA expression and StAR promoter activity in transiently transfected MA-10 cells. Thus, the interaction of quercetin is just opposite to genistein interactions with this particular gene. Therefore, caution should be taken when anybody mixes these two ingredients: diet or supplementations. Quercetin also influences the upregulation of lipoprotein lipase (LPL) gene.

Rosmarinic acid impacts the indoleamine 2,3-dioxygenase (IDO) gene. It suppressed the functional activity of IDO and blocked the IDO-dependent T cell suppression. These findings suggest that rosmarinic acid has potential as a pharmacological and transcriptional inhibitor of IDO. Thus, the natural products mentioned have significant impacts on genes in chromosome 8, particularly on the StAR, LPL, and IDO genes, and the interactions of these selected natural products with other genes of chromosome 8 have not been discussed in this chapter.

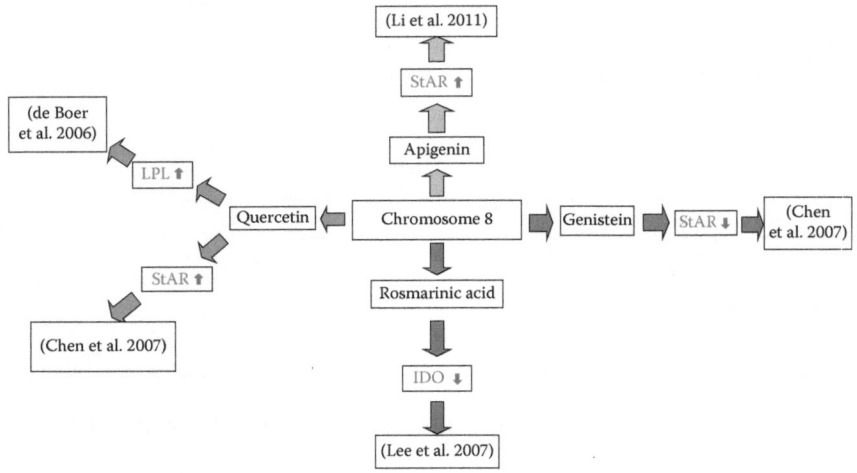

FIGURE 2.8 **(See color insert.)** Natural products interacting with the selected genes of chromosome 8. IDO [IDO and rosmarinic acid]: http://www.ncbi.nlm.nih.gov/pubmed/?term= IDO+AND+Rosmarinic+Acid; LPL [LPL and quercetin]: http://www.ncbi.nlm.nih.gov/ pubmed/?term=LPL+AND+Quercetin; StAR [StAR and apigenin]: http://www.ncbi.nlm.nih. gov/pubmed/?term=StAR+AND+Apigenin; and StAR [StAR and genistein]: http://www.ncbi. nlm.nih.gov/pubmed/?term=StAR+AND+Genistein; and StAR [StAR and quercetin]: http:// www.ncbi.nlm.nih.gov/pubmed/?term=StAR+AND+Quercetin. (Data from Chen, Y. C. et al., *J Endocrinol*, 192, 527–37, 2007; de Boer, V. C. et al., *Cell Mol Life Sci*, 63, 2847–58, 2006; Lee, H. J. et al., *Biochem Pharmacol*, 73, 1412–21, 2007; Li, W. et al., *J Nutr Biochem*, 22, 212–18, 2011.)

2.10   CHROMOSOME 9

Table 2.17 summarizes 24 types of genetic diseases from chromosome 9. Table 2.18 provides the list of genes that are chosen for the natural products interactions studies. As presented in Figure 2.9, the effects of four natural products on the toll-like receptor 4 (TLR4) gene in chromosome 9 have been studied. In one study, ferulic

---

TABLE 2.17
Genetic Diseases Linked to Chromosome 9

1. Acytosiosis
2. ALA-D deficiency porphyria
3. Citrullinemia
4. Ehlers–Danlos syndrome
5. Ehlers–Danlos syndrome, classical type
6. Familial dysautonomia
7. Friedreich ataxia
8. Galactosemia
9. Gorlin syndrome or nevoid basal cell carcinoma syndrome
10. Hereditary hemorrhagic telangiectasia

(Continued)

TABLE 2.17 (*Continued*)

Genetic Diseases Linked to Chromosome 9

11. Lethal congenital contracture syndrome
12. Nail-patella syndrome (NPS)
13. Nonsyndromic deafness
14. Nonsyndromic deafness, autosomal dominant
15. Nonsyndromic deafness, autosomal recessive
16. Obsessive–compulsive disorder
17. Porphyria
18. Primary hyperoxaluria
19. Tangier's disease
20. Tetrasomy 9p
21. Thrombotic thrombocytopenic purpura
22. Trisomy 9
23. Tuberous sclerosis
24. VLDLR-associated cerebellar hypoplasia

TABLE 2.18

Chromosome 9 Selected Genes

PAX5: Paired Box 5

This gene is associated with lymphoplasmacytic lymphoma and diffuses large B cell lymphoma of the central nervous system.

CNTNAP3: Contactin-Associated Protein-Like 3

This gene encodes for a protein that is part of the NCP family of cell-recognition molecules.

STOML2: Stomatin (EPB72)-Like 2

This is associated with acromesomelic dysplasia and waldenstrom macroglobulinemia.

UNC13B: Unc-13 Homolog B (*C. elegans*)

This gene is expressed in the kidney cortical epithelial cells and is upregulated by hyperglycemia.

TPM2: Tropomyosin 2 (Beta)

This gene encodes for beta-tropomyosin, which is part of the actin filament-binding protein family, and it is expressed in slow, type 1 muscle fibers.

XPA: Xeroderma Pigmentosum, Complementation Group A

This gene encodes for a zinc-finger involved in DNA excision repair.

IFNA13: Interferon, Alpha 13

This gene is associated with dengue hemorrhagic fever and hemorrhagic fever.

BAG1: BCL2-Associated Athanogene

This gene encodes for the oncogene BCL2, which is a membrane protein that blocks a step in a pathway leading to apoptosis.

TLR4: Toll-Like Receptor 4

This gene encodes for a protein that is part of TLR family and is important in pathogen recognition and activation of innate immunity.

TMEM2: Transmembrane Protein 2

This gene is associated with benign familial infantile epilepsy and hepatitis B virus infection.

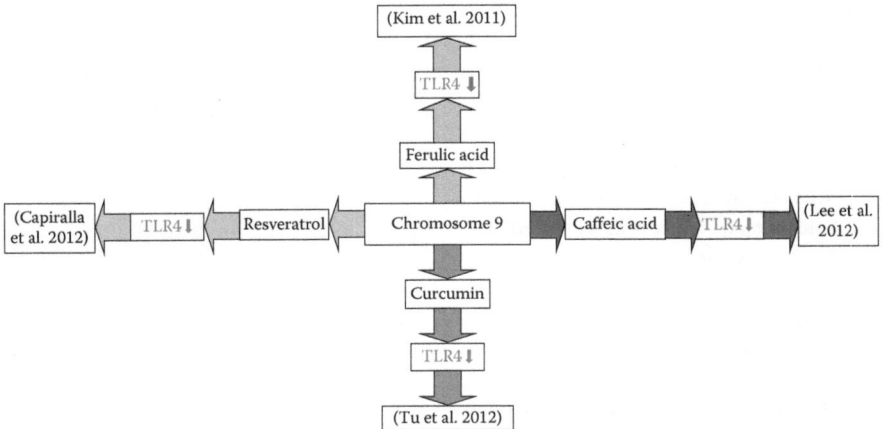

FIGURE 2.9 **(See color insert.)** Natural products interacting with the selected genes of chromosome 9. TLR4 [TLR4 and caffeic acid]: http://www.ncbi.nlm.nih.gov/pubmed/?term =TLR4+AND+Caffeic+Acid; TLR4 [TLR4 and curcumin]: http://www.ncbi.nlm.nih.gov/ pubmed/?term=TLR4+AND+Curcumin; TLR4 [TLR4 and ferulic acid]: http://www.ncbi.nlm. nih.gov/pubmed/?term=TLR4+AND+Ferulic+acid; and TLR4 [TLR4 and resveratrol]: http:// www.ncbi.nlm.nih.gov/pubmed/?term=TLR4+AND+Resveratrol. (Data from Capiralla, H. et al., *J Neurochem*, 120, 461–72, 2012; Kim, H. Y. et al., *Toxicology*, 282, 104–11, 2011; Lee, E. S., *J Agric Food Chem*, 60, 2730–9, 2012; Tu, C. T., *Int Immunopharmacol*, 12, 151–7, 2012.)

acid greatly inhibited TLR4 expression in mice. Another study showed that caffeic acid diminished TLR4 induction. This study suggests caffeic acid may have therapeutic potential in preventing obesity-associated atherosclerosis. Curcumin treatment lowered the expression levels of TLR4 mRNA or protein in liver tissues. This study indicates that the beneficial effect of curcumin might be partly mediated by inhibiting the expression levels of TLR4 in the liver. Finally, resveratrol acted upstream in the activation cascade by interfering with TLR4 oligomerization upon receptor stimulation. This contributes to anti-inflammatory effects against Aβ-triggered microglial activation. Clearly, these natural products have a significant impact on the TLR4 gene in chromosome 9, and the interactions of these selected natural products with other genes of chromosome 9 have not been discussed in this chapter.

2.11   CHROMOSOME 10

Table 2.19 summarizes 21 types of genetic diseases from chromosome 10. Table 2.20 provides the list of genes that are chosen for the natural products interactions studies. As presented in Figure 2.10, four different natural products were studied for their effects on the silent information regulator 1 (SIRT1) gene in chromosome 10. A study conducted on one of these natural products, genistein, revealed that genistein increased SIRT1 expression in renal proximal tubular cells. Quercetin, curcumin, and resveratrol showed that these natural products stimulate SIRT1 both directly and indirectly. This is beneficial for the regulation of calorie restriction, oxidative stress, inflammation, cellular senescence, autophagy and apoptosis, autoimmunity, metabolism, adipogenesis,

TABLE 2.19
Genetic Diseases Linked to Chromosome 10

1. Apert syndrome
2. Beare–Stevenson cutis gyrata syndrome
3. Charcot–Marie–Tooth disease
4. Charcot–Marie–Tooth disease, type 1
5. Charcot–Marie–Tooth disease, type 4
6. Cockayne syndrome
7. Congenital erythropoietic porphyria
8. Cowden syndrome
9. Crouzon syndrome
10. Hirschprung disease
11. Jackson–Weiss syndrome
12. Multiple endocrine neoplasia, type 2
13. Nonsyndromic deafness
14. Nonsyndromic deafness, autosomal recessive
15. Pfeiffer syndrome
16. Porphyria
17. Tetrahydrobiopterin deficiency
18. Thiel–Behnke corneal dystrophy
19. Usher syndrome
20. Usher syndrome, type I
21. Wolman syndrome

TABLE 2.20
Chromosome 10 Selected Genes

SPAG6: Sperm-Associated Antigen 6

This gene is associated with infertility and hydrocephalus.

SIRT1: Sirtuin 1

This gene encodes for a protein that is part of the sirtuin family of proteins and homologs to the yeast SIR2 protein.

NRP1: Neuropilin 1

This gene encodes for two neuropilins, which contain specific protein domains allowing them to participate in several different types of signaling pathways that control cell migration.

SVIL: Supervillin

This gene encodes for a bipartite protein with distinct amino- and carboxy-terminal domains.

ZNF37A: Zinc-Finger Protein 37A

This gene may be involved in transcriptional regulation.

CUL2: Cullin 2

This gene is associated with glomuvenous malformation and von Hippel–Lindau disease.

ITGA8: Integrin, Alpha 8

This gene is part of the genesis of kidney and many other organs by regulating the recruitment of mesenchymal cells into epithelial structures.

(Continued)

TABLE 2.20 *(Continued)*
Chromosome 10 Selected Genes

PHYH: Phytanoyl-Coa 2-Hydroxylase

    This gene is part of the PhyH family, and it encodes for a peroxisomal protein involved in the alpha-oxidation of 3-methyl branched fatty acids.

KIAA1462: KIAA1462

    This gene is associated with morbid obesity and obesity. A coronary artery disease-associated gene is a candidate gene for late onset Alzheimer's disease in APOE carriers.

MYO3A: Myosin IIIA

    This gene encodes for a protein that is part of the myosin superfamily.

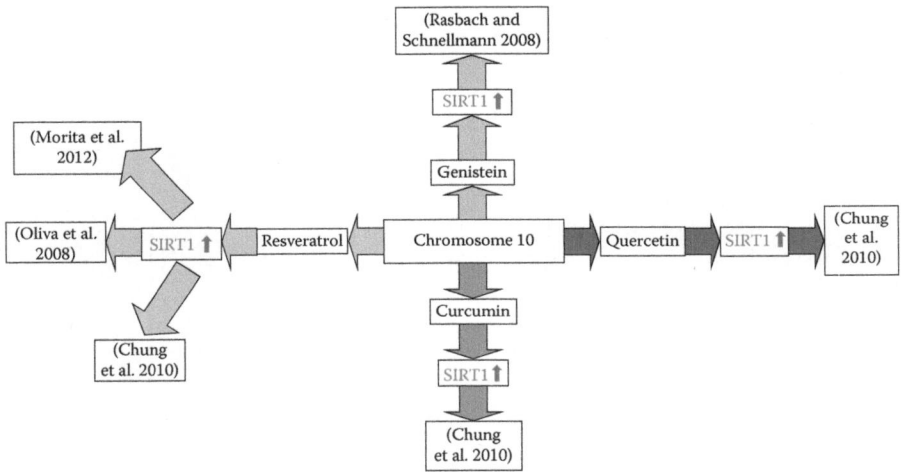

FIGURE 2.10 **(See color insert.)** Natural products interacting with the selected genes of chromosome 10. SIRT1 [SIRT1 and curcumin]: http://www.ncbi.nlm.nih.gov/pubmed/?term=SIRT1+AND+Curcumin; SIRT1 [SIRT1 and genistein]: http://www.ncbi.nlm.nih.gov/pubmed/?term=SIRT1+AND+Genistein; SIRT1 [SIRT1 and quercetin]: http://www.ncbi.nlm.nih.gov/pubmed/?term=SIRT1+AND+Quercetin; and SIRT1 [SIRT1 and resveratrol]: http://www.ncbi.nlm.nih.gov/pubmed/?term=SIRT1+AND+Resveratrol. (Data from Chung, S. et al., *Arch Biochem Biophys*, 501, 79–90, 2010; Morita, Y. et al., *Reprod Biol Endocrinol*, 10, 14, 2012; Oliva, J. et al., *Exp Mol Pathol*, 85, 155–9, 2008; Rasbach, K. A., and R. G. Schnellmann, *J Pharmacol Exp Ther*, 325, 536–43, 2008.)

circadian rhythm, skeletal muscle function, mitochondria biogenesis, and endothelial dysfunction. Furthermore, an additional study demonstrated that resveratrol increases mRNA levels of SIRT1. Yet, another study showed that while ethanol increased the expression of SIRT1, when resveratrol was fed with the ethanol, the expression of SIRT1 did not increase. Therefore, resveratrol prevents the activation of the SIRT1 pathway by ethanol. The natural products mentioned have a significant influence on the SIRT1 gene in chromosome 10, and the interactions of these selected natural products with other genes of chromosome 10 have not been discussed in this chapter.

2.12 CHROMOSOME 11

Table 2.21 summarizes 36 types of genetic diseases from chromosome 11. Table 2.22 provides the list of genes that are chosen for the natural products interactions studies. As presented in Figure 2.11a–d, various natural products impact the genes in chromosome 11. Genistein interacts with multiple genes in chromosome 11. Genistein induced tyrosine hydroxylase (TH) to be upregulated in the mammary glands of rats.

TABLE 2.21
Genetic Diseases Linked to Chromosome 11

1. Autism (neurexin 1)
2. Aniridia
3. Acute intermittent porphyria
4. Albinism
5. Ataxia-telangiectasia
6. Beckwith–Wiedemann syndrome
7. Best's disease
8. Beta-ketothiolase deficiency
9. Beta thalassemia
10. Bladder cancer
11. Breast cancer
12. Carnitine palmitoyltransferase I deficiency
13. Charcot–Marie–Tooth disease
14. Charcot–Marie–Tooth disease, type 4
15. Denys–Drash syndrome
16. Familial Mediterranean fever
17. Hereditary angioedema OMIM: 106100
18. Jacobsen syndrome
19. Jervell and Lange-Nielsen syndrome
20. Meckel syndrome
21. Methemoglobinemia, beta-globin type
22. Mixed lineage leukemia
23. Multiple endocrine neoplasia, type 1
24. Hereditary multiple exostoses
25. Niemann–Pick disease
26. Nonsyndromic deafness
27. Nonsyndromic deafness, autosomal dominant
28. Nonsyndromic deafness, autosomal recessive
29. Porphyria
30. Romano–Ward syndrome
31. Sickle-cell anemia
32. Smith–Lemli–Opitz syndrome
33. Tetrahydrobiopterin deficiency
34. Usher syndrome
35. Usher syndrome, type I
36. WAGR syndrome

TABLE 2.22

Chromosome 11 Selected Genes

CPT1A: Carnitine Palmitoyltransferase 1A

   This gene is associated with carnitine palmitoyltransferase 1A deficiency.

HBB: Hemoglobin, Beta

   This gene is involved in oxygen transport from the lungs to various peripheral tissues.

SBF2: SET-Binding Factor 2

   This gene encodes for a pseudophosphatase and member of the myotubularin-related protein family.

TH: Tyrosine Hydroxylase

   This gene encodes for a protein involved in the conversion of tyrosine to dopamine (DA).

RAG2: Recombination Activating Genes 2

   This gene encodes for a protein involved in the initiation of V (D) J recombination during B and T cell development.

ACAT1: Acetyl-Coenzyme A Acetyltransferase 1

   This gene encodes a mitochondrially localized enzyme that catalyzes the reversible formation of acetoacetyl-CoA from two molecules of acetyl-CoA.

CD81: Cluster of Differentiation 81

   This gene is expressed in hemopoietic, endothelial, and epithelial cells.

WT1: Wilms Tumor 1

   This gene encodes for a transcription factor containing four zinc-finger motifs at the C-terminus and a proline/glutamine-rich DNA-binding domain at the N-terminus.

APOA4: Apolipoprotein A-IV

   This gene contains three exons separated by two introns.

BDNF: Brain-Derived Neurotrophic Factor

   This gene encodes for a member of the nerve growth family, which is induced by cortical neuron and necessary for survival of striatal neurons in the brain.

Genistein is capable of upregulating carnitine palmitoyltransferase 1A (CPT1A) enzyme activity through the upregulation of CPT1A transcription. Furthermore, it was discovered that the co-treatment of L-carnitine and genistein additively increases CPT1A enzyme activity in HepG2 cells. Genistein downregulates RELA in human myeloma cells, resulting in the suppression of proliferation and the induction of apoptosis. Genistein also greatly increases the expression of BDNF mRNA and protein levels.

Curcumin weakens the Wilms' tumor gene (WT)-1 autoregulatory function through the inhibition of protein kinase C (PKC)-α signaling in K562 cells. Curcumin causes a downregulation of IL18 expression 9. In addition, curcumin substantially decreases the brain BDNF levels in corticosterone (CORT)-treated rats. Curcumin also activates the CASP3 gene in LoVo cells. Additionally, research shows that curcumin can reverse chemoresistance by downregulating RELA through IL18 in human gastric cancer cells.

Ferulic acid interacts with several genes in chromosome 11. It could be used as whitening agent, which suppresses melanogenesis by inhibiting the TH gene activity in an indirect manner. Oral administration of ferulic acid increased BDNF mRNA levels in the hippocampus of CORT-treated mice. Further study found that ferulic acid considerably reduced CASP3 expression in kidney cells.

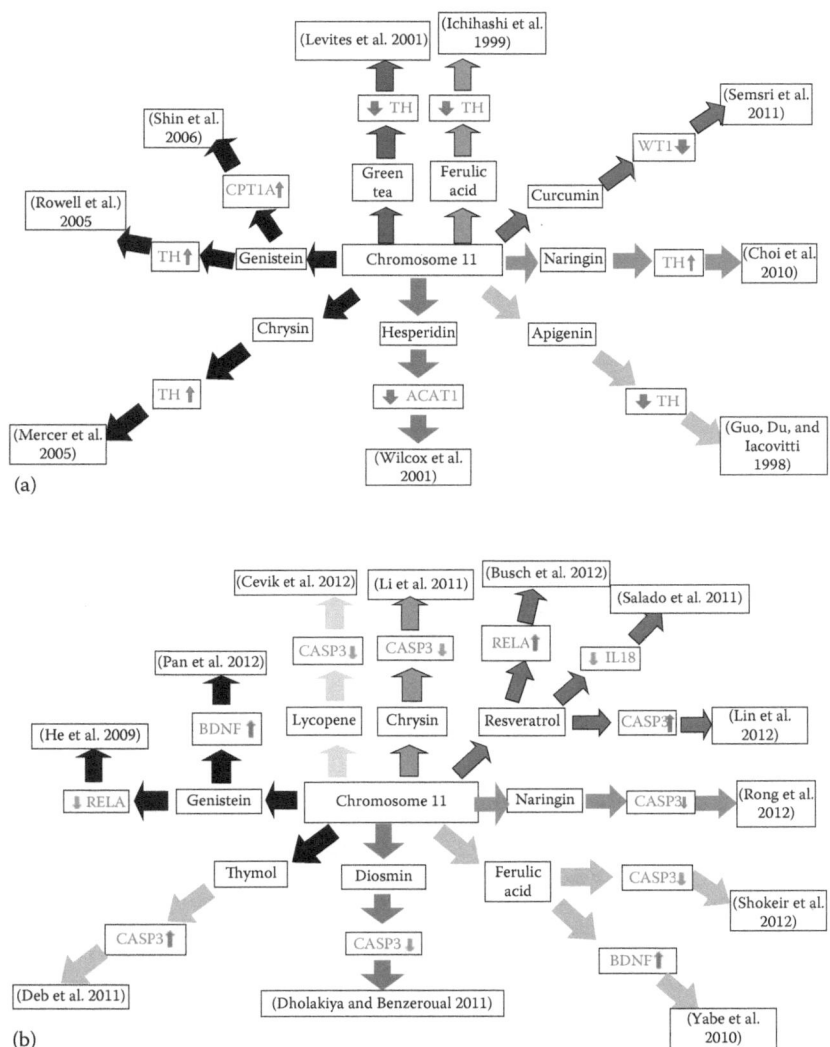

FIGURE 2.11 **(See color insert.)** Natural products interacting with the selected genes of chromosome 11. (a) ACAT1 [ACAT1 and hesperidin]: http://www.ncbi.nlm.nih.gov/pubmed/?term =ACAT1+AND+Hesperidin; CPT1A [CPT1A and genistein]: http://www.ncbi.nlm.nih.gov/pubmed/?term=CPT1A+AND+Genistein; TH [TH and apigenin]: http://www.ncbi.nlm.nih.gov/pubmed/?term=Tyrosine+hydroxylase+AND+Apigenin; TH [TH and ferulic acid]: http://www.ncbi.nlm.nih.gov/pubmed/?term=Tyrosine+hydroxylase+AND+ferulic+acid; TH [TH and chrysin]: http://www.ncbi.nlm.nih.gov/pubmed/?term=Tyrosine+hydroxylase+AND +Chrysin; TH [TH and genistein]: http://www.ncbi.nlm.nih.gov/pubmed/16317154; TH [TH and green tea]: http://www.ncbi.nlm.nih.gov/pubmed/?term=Tyrosine+hydroxylase+AND+ green+tea; TH [TH and naringin]: http://www.ncbi.nlm.nih.gov/pubmed/?term=Tyrosine+ hydroxylase+AND+Naringin; and WT1 [WT1 and curcumin]: http://www.ncbi.nlm.nih.gov/pubmed/?term=WT1+AND+curcumin. (Data from Choi, B. S. et al., *Neurochem Res*, 35, 1269–80, 2010; Guo, Z. *J Neurosci*, 18, 8163–74, 1998; *(Continued)*

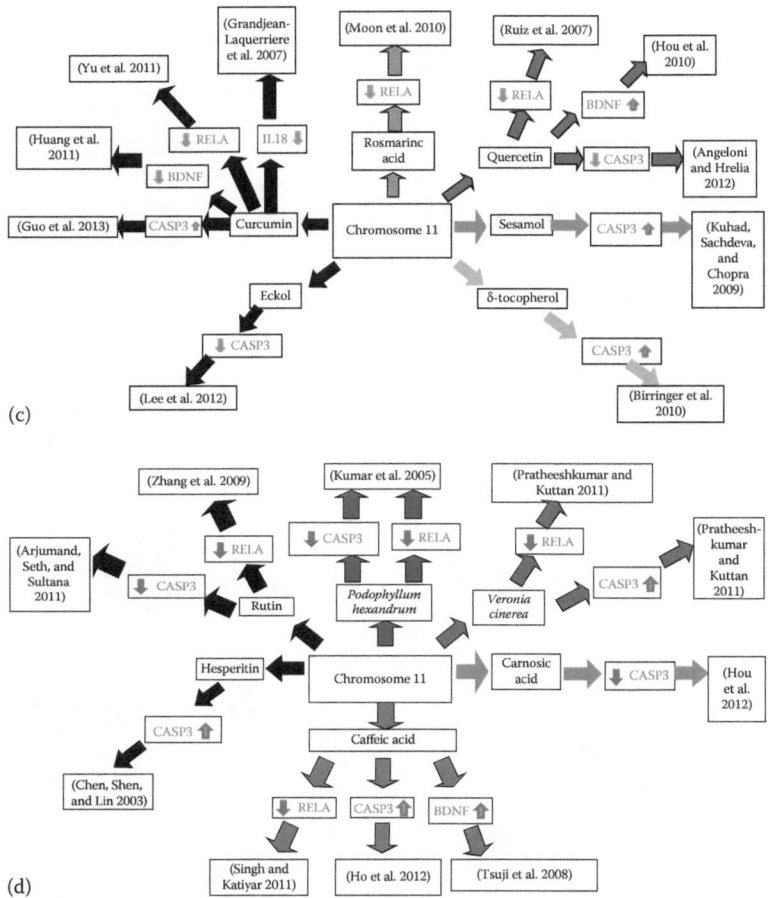

(c)

(d)

FIGURE 2.11 (Continued) Ichihashi, M. et al., *Anticancer Res*, 19, 3769–74, 1999; Rowell, C. et al., *J Nutr*, 135:2953S-2959S, 2005; Levites, Y. et al., *J Neurochem*, 78, 1073–82, 2001; Mercer, L. D. et al., *Biochem Pharmacol*, 69, 339–45, 2005; Semsri, S. et al., *FEBS Lett*, 585, 2235–42, 2011; Shin, E. S. et al., *Eur J Nutr*, 45, 159–64, 2006; Wilcox, L. J. et al., *J Lipid Res*, 42, 725–34, 2001.) (b) BDNF [BDNF and ferulic acid]: http://www.ncbi.nlm.nih. gov/pubmed/?term=BDNF+AND+Ferulic+acid; BDNF [BDNF and genestein]: http://www. ncbi.nlm.nih.gov/pubmed/?term=BDNF+AND+genestein; CASP3 [CASP3 and chrysin]: http://www.ncbi.nlm.nih.gov/pubmed/?term=CASP3+AND+Chrysin; CASP3 [CASP3 and diosmin]: https://www.ncbi.nlm.nih.gov/pubmed/?term=Diosmin+AND+CASP3; CASP3 [CASP3 ferulic acid] https://www.ncbi.nlm.nih.gov/pubmed/?term=Ferulic+acid+ AND+CASP3; CASP3 [CASP3 and lycopene]: http://www.ncbi.nlm.nih.gov/pubmed/?term= CASP3+AND+Lycopene; CASP3 [CASP3 and naringin]: https://www.ncbi.nlm.nih.gov/ pubmed/? term=Naringin+AND+CASP3; CASP3 [CASP3 and resveratrol]: https://www.ncbi. nlm.nih.gov/pubmed/?term=Resveratrol+AND+CASP3; CASP3 [CASP3 and thymol]: https:// www.ncbi.nlm.nih.gov/pubmed/?term=Thymol+AND+CASP3; IL18 [IL18 and resveratrol]: http://www.ncbi.nlm.nih.gov/pubmed/?term=IL18+AND+Resveratrol; RELA [RELA and genistein]: http://www.ncbi.nlm.nih.gov/pubmed/?term= RELA+AND+Genistein;   (*Continued*)

FIGURE 2.11 (Continued) and RELA [RELA and resveratrol]: http://www.ncbi.nlm.nih. gov/pubmed/22936809. (Data from Busch, F. et al., *J Biol Chem*, 287, 25770–81, 2012; Cevik, O. et al., *Burns*, 38, 861–71, 2012; Deb, D. D. et al., *Chem Biol Interact*, 193, 97–106, 2011; Dholakiya, S. L., and K. E. Benzeroual, *Toxicol In Vitro*, 25, 1039–44, 2011; He, H. et al., *Phytother Res*, 23, 868–73, 2009; Li, X. et al., *Toxicol In Vitro*, 25, 630–5, 2011; Lin, X. et al., *Int J Urol*, 19, 757–64, 2012; Pan, M. et al., *J Nutr Health Aging*, 16, 389–94, 2012; Rong, W. et al., *Neurochem Res*, 37, 1615–23, 2012; Salado, C. et al., *J Transl Med*, 9, 59, 2011; Shokeir, A. A. et al., *BJU Int*, 110, 904–11, 2012; Yabe, T. et al., *Neuroscience*, 165, 515–24, 2010.) (c) BDNF [BDNF and curcumin]: https://www.ncbi.nlm.nih.gov/ pubmed/?term=curcumin+AND+BDNF; BDNF [BDNF and quercetin]: https://www.ncbi. nlm.nih.gov/pubmed/?term=Quercetin+AND+BDNF; CASP3 [CASP3 and curcumin]: https:// www.ncbi.nlm.nih.gov/pubmed/?term=Curcumin+AND+CASP3; CASP3 [CASP3 and eckol]: https://www.ncbi.nlm.nih.gov/pubmed/?term=Eckol+AND+CASP3; CASP3 [CASP3 and quercetin]: https://www.ncbi.nlm.nih.gov/pubmed/?term=Quercetin+AND+CASP3; CASP3 [CASP3 and sesamol]: https://www.ncbi.nlm.nih.gov/pubmed/?term=Sesamol+AND+CASP3; CASP3 [CASP3 and δ-tocopherol]: https://www.ncbi.nlm.nih.gov/pubmed/?term=%CE%B4-tocopherol+AND+CASP3; IL18 [IL18 and curcumin]: https://www.ncbi.nlm.nih.gov/pubmed /?term=curcumin++AND+IL18; RELA [RELA and curcumin]: https://www.ncbi.nlm.nih.gov/ pubmed/?term=curcumin+AND+RELA; RELA [RELA and quercetin]: https://www.ncbi.nlm. nih.gov/pubmed/?term=Quercetin+AND+RELA; RELA [RELA and rosmarinc acid]: https:// www.ncbi.nlm.nih.gov/pubmed/?term=Rosmarinc+acid+AND+RELA. (Data from Angeloni, C., and S. Hrelia, *Oxid Med Cell Longev*, 2012, 837104, 2012; Birringer, M. et al., *Free Radic Biol Med*, 49, 1315–22, 2010; Grandjean-Laquerriere, A., *Cytokine*, 37, 76–83, 2007; Guo, L. D., *Phytother Res*, 27, 422–30, 2013; Han, S. S. et al., *J Cell Biochem*, 93, 257–70, 2004; Hou, Y. et al., *Neuropharmacology*, 58, 911–20, 2010; Huang, Z. et al., *Neurosci Lett*, 493, 145–8, 2011; Kuhad, A. et al., *J Agric Food Chem*, 57, 6123–8, 2009; Lee, M. S. et al., *J Agric Food Chem*, 60, 5340–9, 2012; Moon, D. O., *Cancer Lett*, 288, 183–91, 2010; Ruiz, P. A. et al., *J Nutr*, 137, 1208–15, 2007; Tsuji, M. et al., *Nihon Shinkei Seishin Yakurigaku Zasshi*, 28, 159–67, 2008; Yu, L. L. et al., *Oncol Rep*, 26, 1197–203, 2011.) (d) BDNF [BDNF and caffeic acid]: https:// www.ncbi.nlm.nih.gov/pubmed/?term=Caffeic+acid+AND+BDNF; CASP3 [CASP3 and caffeic acid]: https://www.ncbi.nlm.nih.gov/pubmed/?term=Caffeic+acid+AND+CASP3; CASP3 [CASP3 [CASP3 and carnosic acid]: https://www.ncbi.nlm.nih.gov/pubmed/?term=Carnosic+ acid+AND+CASP3; CASP3 [CASP3 and hesperitin]: https://www.ncbi.nlm.nih.gov/pubmed/? term=Hesperitin+AND+CASP3; CASP3 [CASP3 and *Podophyllum hexandrum*]: https://www. ncbi.nlm.nih.gov/pubmed/?term=Podophyllum+hexandrum+AND+CASP3; CASP3 [CASP3 and rutin]: https://www.ncbi.nlm.nih.gov/pubmed/?term=Rutin+AND+CASP3; CASP3 [CASP3 and *Veronia cinerea*]: https://www.ncbi.nlm.nih.gov/pubmed/?term=Veronia+cinerea+ AND+CASP3; RELA [RELA and caffeic acid]: https://www.ncbi.nlm.nih.gov/pubmed/?term= Caffeic+acid+AND+RELA; RELA [RELA and *Podophyllum hexandrum*]: https://www.ncbi. nlm.nih.gov/pubmed/?term=Podophyllum+hexandrum+AND+CASP3; RELA [RELA and rutin]: https://www.ncbi.nlm.nih.gov/pubmed/?term=Rutin+AND+RELA; and RELA [RELA and *Veronia cinerea*]: https://www.ncbi.nlm.nih.gov/pubmed/?term=Veronia+cinerea+AND+ CASP3. (Data from Arjumand, W. et al., *Food Chem Toxicol*, 49, 2013–21, 2011; Chen, Y. C. et al., *Biochem Pharmacol*, 66, 1139–50, 2003; Ho, H. C. et al., *J Vasc Res*, 49, 24–32, 2012; Hou, C. W. et al., *Nutr Neurosci*, 2012; Kumar, R. et al., *Environ Toxicol Pharmacol*, 20, 326–34, 2005; Pratheeshkumar, P., and G. Kuttan, *J Environ Pathol Toxicol Oncol*, 30, 139–51, 2011; Singh, T., and S. K. Katiyar, *PLoS One*, 6, e25224, 2011; Tsuji, M. et al., *Nihon Shinkei Seishin Yakurigaku Zasshi*, 28, 159–67, 2008; Zhang, Z. F. et al., *J Agric Food Chem*, 57, 7731–6, 2009.)

Quercetin influences various genes in chromosome 11. For example, a study demonstrated that quercetin has inhibitory effects on phospho-RelA recruitment to the IP-10. An additional study showed that quercetin pretreatment significantly counteracted apoptosis cell death as measured by the immunoblotting of the cleaved caspase (CASP) 3 and CASP3 activity. Furthermore, quercetin increases the expression of the BDNF gene in neurons and hippocampus of double TgAD mice.

Caffiec acid interacts with CASP3, BDNF, and RELA genes. CAPE instigated the upregulation of CASP3; this suggests that caffeic acid precipitates the mitochondrion-dependent apoptotic-signaling pathway. Caffeic acid attenuates the downregulation of BDNF transcription. Also, caffeic acid inhibits the RELA gene. Thus, caffeic acid and curcumin have similar interacting effects on BDNF and RELA genes.

Resveratrol influences RELA, IL18, and CASP3 genes. Resveratrol causes the upregulation of RELA. Resveratrol inhibits IL18 activity. Consequently, resveratrol significantly inhibited hepatic retention and metastatic growth of melanoma cells. Moreover, resveratrol drastically promoted the activation of CASP3 in bladder cancer cells. This suggests that resveratrol proficiently triggers apoptosis in bladder cancer cells. Thus, resveratrol has pharmacological potential in the treatment of bladder cancer.

Many other natural products affect different genes in chromosome 11, particularly the CASP3 gene. Hesperidin increased CASP3 activity. Chrysin could increase TRAIL-induced degradation of CASP3. (TRAIL is known as Apo-2L and is a member of the TNF ligand family in apoptosis.) Lycopene is used for the treatment against oxidative injury in rats with thermal trauma. Treatment with lycopene inhibited the CASP3 gene. This suggests that lycopene possesses anti-inflammatory, antiapoptotic, and antioxidant effects that prevent burn-induced oxidative damage in remote organs. A different study revealed that naringin treatment greatly reduced the enzyme activity of CASP3 and decreased the number of apoptotic cells after a spinal cord injury. These findings suggest that naringin treatment could improve functional recovery after a spinal cord injury.

Another study revealed that diosmin inhibited LPS-induced CASP3 activation. This suggests that diosmin has anti-apoptotic effects, and caution should be exercised while treating patients with caspase-activating chemotherapies. Thymol increases the CASP3 activation. Sesamol increases the activity of CASP3. Mice that were treated with *Podophyllum hexandrum* exhibited a decrease in CASP3. Moreover, *P. hexandrum* inhibited RELA in these mice. δ-Tocopherol increased CASP3 activation. Carnosic acid inhibited CASP3 activation. Eckol inhibited the expression of CASP3. A different study showed that *V. cinerea* induces apoptosis through the activation of p53-induced, CASP3-mediated proapoptotic signaling and the suppression of RELA-induced, bcl-2-mediated survival signaling. In contrary, rutin has an inhibitory effect on CASP3. An additional study showed that troxerutin, a trihydroxyethylated derivative of rutin, inhibited the upregulation of the expression of RELA. Thus, the interactions of rutin and its derivatives are quite different from caffeic acid and curcumin on the gene RELA.

Apigenin eliminated TH expression and the associated AP-1 changes. Tripterygium regelii extract (TRE) containing naringin increased TH in SH-SY5Y cells. Chrysin protected mesencephalic cultures from injury by MPP+, which was displayed by DNA fragmentation studies and TH immunocytochemistry of DA neurones to occur by apoptosis. A different study revealed that hesperidin reduced activities of ACAT1 gene. This mediated the decrease in the availability of lipids for assembly of apoB-containing lipoproteins. Rosmarinic acid inhibits RELA activation. L-Ascorbic acid causes the downregulation of NF-kappaB activity. This contributes to the RELA antitumor activity. These numerous natural products clearly have a significant impact on various genes in chromosome 11, and the interactions of these selected natural products with other genes of chromosome 11 have not been discussed in this chapter.

2.13   CHROMOSOME 12

Table 2.23 summarizes 26 types of genetic diseases from chromosome 12. Table 2.24 provides the list of genes that are chosen for the natural products interactions studies. As presented in Figure 2.12, there are several natural products that affect genes in chromosome 12. One of the natural products is chrysin, which affects the cyclin-dependent kinase 2 (CDK2). Chrysin reduced the CDK2 in a dose-depending manner. Hesperidin inhibited the expression of CDK2. Quercetin affects the genes CDK2, ubiquitin C (UBC), and KRAS. Quercetin induces the G0/G1 phase arrest by decreasing the levels of CDK2. Quercetin also reduces the expression on the polyubiquitin gene (UBC) in

TABLE 2.23

Genetic Diseases Linked to Chromosome 12

1. Achondrogenesis, type 2
2. Collagenopathy, types II and XI
3. Cornea plana 2
4. Episodic ataxia
5. Hereditary hemorrhagic telangiectasia
6. Hypochondrogenesis
7. Ichthyosis bullosa of Siemens
8. Kniest dysplasia
9. Maturity onset diabetes of the young, type 3
10. Methylmalonic acidemia
11. Narcolepsy
12. Nonsyndromic deafness
13. Nonsyndromic deafness, autosomal dominant
14. Noonan syndrome
15. Parkinson's disease
16. Pallister–Killian syndrome (tetrasomy 12p)
17. Phenylketonuria
18. Spondyloepimetaphyseal dysplasia, Strudwick type
19. Spondyloepiphyseal dysplasia congenita
20. Spondyloperipheral dysplasia
21. Stickler syndrome
22. Stickler syndrome, COL2A1
23. Stuttering
24. Triose phosphate isomerase deficiency
25. Tyrosinemia
26. Von Willebrand disease

TABLE 2.24
Chromosome 12 Selected Genes

COL2A1: Collagen, Type II, Alpha 1

This gene encodes for a fibrillar collagen found in cartilage and the vitreous humor of the eye.

KERA: Keratocan

This gene encodes for a keratin sulfate proteoglycan involved in corneal transparency.

KCNA1: Potassium Voltage-Gated Channel

This gene encodes for voltage-gated channel potassium that is phylogenetically related to the Drosophila Shake channel.

PAH: Phenylalanine Hydroxylase

This gene encodes for the enzyme that is rate-limiting step in phenylalanine catabolism.

PPP1R12A: Protein Phosphatase 1

This gene provides instruction for making the myosin phosphatase target subunit 1.

KRAS: Kirsten Rat Sarcoma Viral Oncogene Homolog

This gene encodes for a protein that is part of the small GTPase superfamily.

ACVRL1: Activin A Receptor Type II-Like 1

This gene encodes for the activin receptor-like kinase 1 protein that is found on the surface of cells.

MYO1A: Myosin IA

This gene encodes for a protein that represents an unconventional myosin, which is different from the skeletal muscle myosin-1.

LRRK2: Leucine-Rich Repeat Kinase 2

This gene is active in the brain and other tissues throughout the body. This gene is part of the family genes of PARK.

HPD: 4-Hydroxyphenylpyruvate Dioxygenase

This gene encodes for an enzyme in the catabolic pathway of tyrosine.

---

the rat liver. In addition, quercetin reduces the phosphorylation of ERK and (where KRAS gene is mutated in HCT15 cell) promotes the cell death.

Naringin downregulates the CDK2 gene. Lycopene also reduces CDK2 activity. Lycopene increases KRAS gene-mutated colon cancer cell death. Ascorbic acid has an effect on the keratin sulfate proteoglycans (KERA) gene. It was found that insulin and ascorbate increased the accumulation of the keratin sulfate proteoglycans (KERA).

Curcumin interacts with abattery of genes such as UBC, CDK2, and leucine-rich repeat kinase 2 (LRRK2) in chromosome 12. Curcumin upregulated 22 genes, and UBC was one of those genes. The next study showed that curcumin induces apoptosis in breast cancer cells by regulating different signaling pathways and showing potential in preventing and treating cancer through the inhibition of the expression of CDK2. Also, it was found that in rat mesencephalic cells, curcumin induced the expression of LRRK2 mRNA and protein in a time-dependent manner.

Resveratrol stops the growth and development of pancreatic cancer in KRAS (G12D) mice. This study suggests that in KRAS (G12D) transgenic mice, resveratrol inhibits pancreatic cancer stem cell characteristics by preventing pluripotency maintaining factors and epithelial-mesenchymal transition. Thus, there are many natural products that affect the genes in chromosome 12, and the interactions of

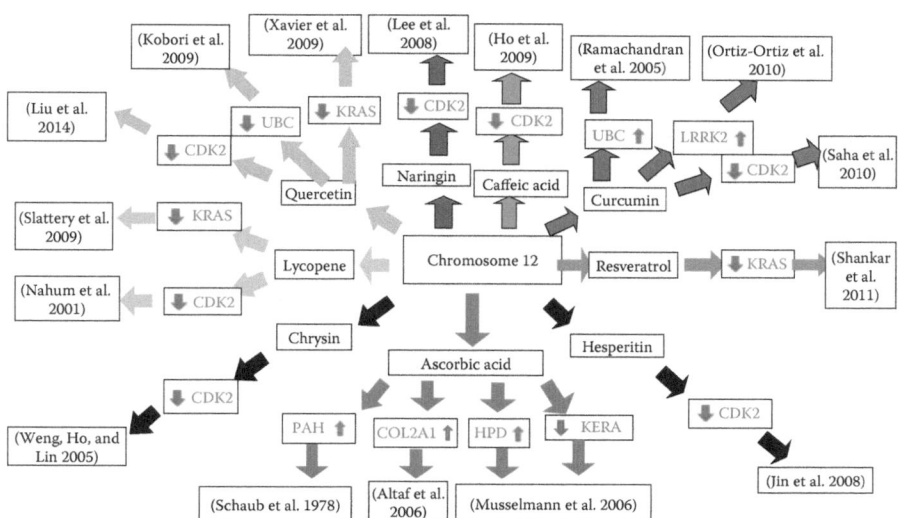

FIGURE 2.12   (See color insert.) Natural products interacting with the selected genes of chromosome 12. CDK2 [CDK2 and caffeic acid]: https://www.ncbi.nlm.nih.gov/pubmed/?term =Caffeic+acid+AND+CDK2; CDK2 [CDK2 and chrysin]: https://www.ncbi.nlm.nih.gov/ pubmed/?term=Chrysin+AND+CDK2; CDK2 [CDK2 and curcumin]: https://www.ncbi.nlm. nih.gov/pubmed/?term=Curcumin+AND+CDK2; CDK2 [CDK2 and hesperitin]: https://www. ncbi.nlm.nih.gov/pubmed/?term=Hesperitin+AND+CDK2; CDK2 [CDK2 and lycopene]: https://www.ncbi.nlm.nih.gov/pubmed/?term=Lycopene+AND+CDK2; CDK2 [CDK2 and naringin]: https://www.ncbi.nlm.nih.gov/pubmed/?term=Naringin+AND+CDK2; CDK2 [CDK2 and quercetin]: https://www.ncbi.nlm.nih.gov/pubmed/?term=Quercetin+AND+ CDK2; COL2A1 [COL2A1 and ascorbic acid]: https://www.ncbi.nlm.nih.gov/pubmed/?term= Ascorbic+Acid+AND+COL2A1; HPD [HPD and ascorbic acid]: https://www.ncbi.nlm.nih. gov/pubmed/?term=Ascorbic+Acid+AND+HPD; KERA [KERA and ascorbic acid]: https:// www.ncbi.nlm.nih.gov/pubmed/?term=Ascorbic+Acid+AND+KERA; KRAS [KRAS and lycopene]: https://www.ncbi.nlm.nih.gov/pubmed/?term=Lycopene+AND+KRAS; KRAS [KRAS and quercetin]: https://www.ncbi.nlm.nih.gov/pubmed/?term=Quercetin+AND+ KRAS; KRAS [KRAS and resveratrol]: https://www.ncbi.nlm.nih.gov/pubmed/?term=Resvera trol+AND+KRAS; LRRK2 [LRRK2 and curcumin]: https://www.ncbi.nlm.nih.gov/pubmed/? term=Curcumin+AND+LRRK2; PAH [PAH and ascorbic acid]: https://www.ncbi.nlm.nih. gov/pubmed/?term=Ascorbic+Acid+AND+PAH; UBC [UBC and curcumin]: https://www. ncbi.nlm.nih.gov/pubmed/?term=Curcumin+AND+UBC; and UBC [UBC and quercetin]: https:// www.ncbi.nlm.nih.gov/pubmed/?term=Quercetin+AND+UBC. (Data from Altaf, F. M. et al., *Eur Cell Mater*, 12, 64–9; discussion 69–70, 2006; Ho, H. C. et al., *Cell Mol Biol [Noisy-le-grand]*, 55:OL1161–7, 2009; Jin, Y. R. et al., *J Cell Biochem*, 104, 1–14, 2008; Kobori, M. et al., *Mol Nutr Food Res*, 53, 859–68, 2009; Lee, E. J. et al., *Food Chem Toxicol*, 46, 3800–7, 2008; Liu, K. C. et al., *Environ Toxicol*, 29, 428–39, 2014; Musselmann, K. et al., *Invest Ophthalmol Vis Sci*, 47, 5260–6, 2006; Nahum, A. et al., *Oncogene*, 20, 3428–36, 2001; Ortiz-Ortiz, M. A. et al., *Neurosci Lett*, 468, 120–4, 2010; Ramachandran, C. et al., *Anticancer Res*, 25, 3293–302, 2005; Saha, A. et al., *Biol Pharm Bull*, 33, 1291–9, 2010; Schaub, J. et al., *Arch Dis Child*, 53, 674–6, 1978; Shankar, S. et al., *PLoS One*, 6, e16530, 2011; Slattery, M. L. et al., *Int J Cancer*, 125, 1698–704, 2009; Weng, M. S. et al., *Biochem Pharmacol*, 69, 1815–27, 2005; Xavier, C. P. et al., *Cancer Lett*, 281, 162–70, 2009.)

these selected natural products with other genes of chromosome 12 have not been discussed in this chapter.

2.14 CHROMOSOME 13

Table 2.25 summarizes 13 types of genetic diseases from chromosome 13. Table 2.26 provides the list of genes that are chosen for the natural products interactions studies. As presented in Figure 2.13, several natural products affect the Rb1 gene in

TABLE 2.25
Genetic Diseases Linked to Chromosome 13

1. Bladder cancer
2. Breast cancer
3. Heterochromia
4. Hirschsprung's disease
5. Maturity onset diabetes of the young, type 4
6. Nonsyndromic deafness
7. Nonsyndromic deafness, autosomal dominant
8. Nonsyndromic deafness, autosomal recessive
9. Propionic acidemia
10. Retinoblastoma
11. Waardenburg syndrome
12. Wilson's disease
13. Patau syndrome

TABLE 2.26
Chromosome 13 Selected Genes

ATP7B: ATPase, Cu^{++} Transporting Beta Polypeptide
 This gene encodes for a protein that is part of the P-type cation transport ATPase family.
BRCA2: Breast Cancer 2, Early Onset
 This gene increases lifetime risk of breast and ovarian cancers.
EDNRB: Endothelin Receptor, Type B
 This gene encodes for a G-protein-coupled receptor that activates a phosphatidylinositol-calcium second messenger system.
GJB2: Gap Junction Protein, Beta 2
 This gene encodes for a protein that is part of the connexin protein family.
SLITRK1: SLIT And NTRK-Like Family, Member 1
 This gene encodes for proteins that are found in the brain and are important for the growth and development of nerve cells.
HTR2A: 5-Hydroxytryptamine Receptor 2A, G Protein-Coupled
 This gene encodes for one of the receptors for serotonin.
PCCA: Propionyl Coenzyme A Carboxylase, Alpha Polypeptide
 This gene encodes for the biotin-binding region of the enzyme propionyl-CoA carboxylase.

(Continued)

TABLE 2.26 (*Continued*)
Chromosome 13 Selected Genes

RB1: Retinoblastoma 1

This gene encodes for a negative regulator of the cell cycle and is the first tumor suppressor gene discovered.

FLT1: Fms-Related Tyrosine Kinase 1

This gene encodes for the vascular endothelial growth factor receptor family.

SOX21. Transcription Factor SOX-21

This gene encodes for a protein that is part of DNA-binding protein containing a 79-amino acid HMG domain.

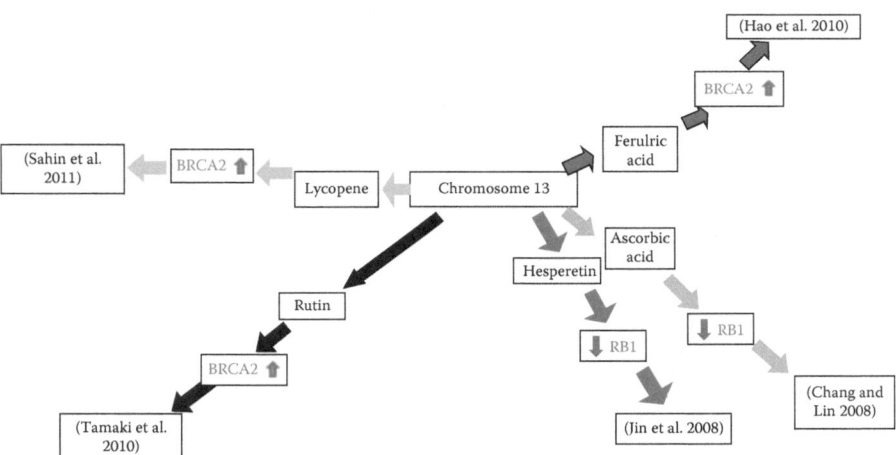

FIGURE 2.13 (See color insert.) Natural products interacting with the selected genes of chromosome 13. BRCA2 [BRCA2 and ferulric acid]: https://www.ncbi.nlm.nih.gov/pubmed/?term=Ferulric+Acid+AND+BRCA2; BRCA2 [BRCA2 and lycopene]: https://www.ncbi.nlm.nih.gov/pubmed/?term=Lycopene+AND+BRCA2; BRCA2 [BRCA2 and rutin]: https://www.ncbi.nlm.nih.gov/pubmed/?term=Rutin+AND+BRCA2; and RB1 [RB1 and ascorbic acid]: https://www.ncbi.nlm.nih.gov/pubmed/?term=Ascorbic+Acid+AND+RB1. (Data from Chang, Y. C., and P. Lin, *Toxicol Appl Pharmacol*, 228, 76–83, 2008; Hao, Q. et al., *Zhongguo Zhong Yao Za Zhi*, 35, 2752–5, 2010; Jin, Y. R. et al., *J Cell Biochem*, 104, 1–14, 2008; Sahin, K. et al., *Nutr Cancer*, 63, 1279–86, 2011; Tamaki, H. et al., *Drug Metab Pharmacokinet*, 25, 170–9, 2010.)

chromosome 13. Genistein inhibits the Rb1 phosphorylation. Ferulic acid also inhibits the phosphorylation of Rb1. Likewise, hesperidin inhibits the phosphorylation of Rb1. Lycopene also affects the Rb1 gene. A study demonstrated that LNCaP and PC3 cells treated with the lycopene-based agents experienced suppressed Rb phosphorylation. In contrast, ascorbic acid enhanced Rb phosphorylation in 1-μM tt-DDE-treated cells. Another study revealed that apigenin caused a decrease in total Rb protein in LNCaP and PC3 cells. Thus, various natural products exhibit a significant contradictory influence on the Rb1 gene in chromosome 13, and always caution should be exercised when using these natural products with or without chemotherapies.

Another very important gene in this chromosome is BRAC2. There are few natural products that interact with this tumor suppressor gene. Thymol, rutin, lycopene, narigin, green tea, carnosic acid, and ferulric acid have profound increase in BRCA2 gene activity and exhibit the anticancer property in the experimental settings. The interactions of these selected natural products with other genes of chromosome 13 have not been discussed yet.

2.15 CHROMOSOME 14

Table 2.27 summarizes 14 types of genetic diseases from chromosome 14. Table 2.28 provides the list of genes that are chosen for the natural products interactions studies. As presented in Figure 2.14, chromosome 14 has only two genes affected by the natural products. Resveratrol is one natural product that affects the GCH1 gene. Resveratrol upregulates the GCH1 gene in human endothelial cells. Hesperindin acts on the

---

TABLE 2.27
Genetic Diseases Linked to Chromosome 14

1. Alpha-1 antitrypsin deficiency
2. Alzheimer's disease
3. Alzheimer's disease, type 3
4. Congenital hypothyroidism
5. DA-responsive dystonia
6. Krabbe disease
7. Machado–Joseph disease
8. Multiple myeloma
9. Niemann–Pick disease
10. Nonsyndromic deafness
11. Nonsyndromic deafness, autosomal dominant
12. Sensenbrenner syndrome
13. Tetrahydrobiopterin deficiency
14. Uniparental disomy (UPD) 14

TABLE 2.28
Chromosome 14 Selected Genes

COCH: Coagulation Factor C Homolog, Cochlin (Limulus poluphemus)
 This gene encodes for the cochlin protein, which is part of the inner ear cochlea and the vestibular system.
GALC: Galactosylceramidase
 This gene encodes for the enzyme that uses water molecules to break down galactolipids, which are found mostly in the brain and kidneys.
IFT43: Intraflagellar Transport 43
 This gene encodes for a subunit of the intraflagellar transport complex A.

(Continued)

TABLE 2.28 (*Continued*)
Chromosome 14 Selected Genes

GCH1: GTP Cyclohydrolase 1

    This gene encodes for the enzyme involved in the first of three steps in producing tetrahydrobiopterin.

NPC2: Niemann–Pick Disease, Type C2

    This gene encodes for the protein that binds to cholesterol and is associated with Niemann–Pick disease.

PSEN1: Presenilin 1

    This gene is associated with Alzheimer's disease and psen1-related dilated cardiomyopathy.

TSHR: Thyroid Stimulation Hormone Receptor

    This gene encodes for the protein that is a receptor for thyrothropin and thyrostimulin.

IGH: Immunoglobulin Heavy Chain Locus

    This gene encodes for immunoglobilins that recognize foreign antigens and start an immune response.

ATXN3: Ataxin-3

    This gene encodes for a protein that is involved in the ubiquitin-proteasome system.

MYH7: Myosin Heavy Chain Beta Isoform

    This gene encodes for a protein found in the heart muscle and in type I skeletal muscle fibers.

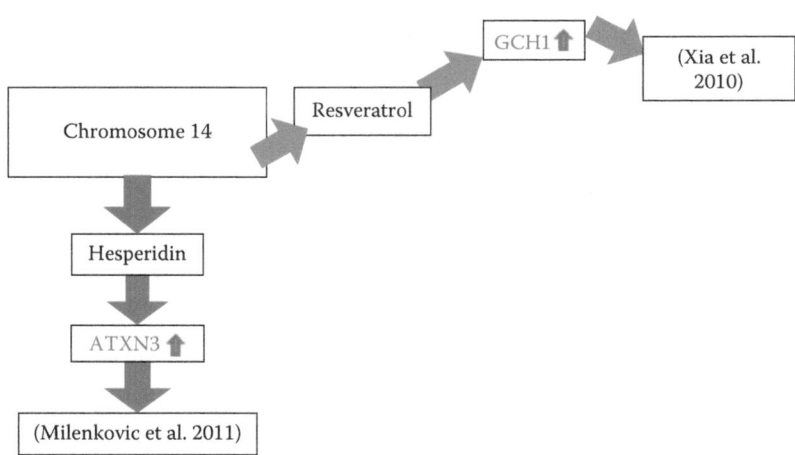

FIGURE 2.14 (See color insert.) Natural products interacting with the selected genes of chromosome 14. ATXN3 [ATXN3 and hesperidin]: https://www.ncbi.nlm.nih.gov/pubmed/22110589; and GCH1 [GCH1 and resveratrol]: https://www.ncbi.nlm.nih.gov/pubmed/?term=Resveratrol+AND+GCH1. (Data from Milenkovic, D. et al., *PLoS One*, 6, e26669, 2011; Xia, N. et al., *J Pharmacol Exp Ther*, 335, 149–54, 2010.)

Ataxin-3 (ATXN3) gene that encodes a protein involved in the ubiquitin-proteasome system that is involved in the intracellular proteolytic enzyme machinery. In addition, the compound has a profound effect on expressions of several other genes from different chromosomes. Additional research needs to be done to find out more about chromosome 14 and its genes interaction with natural products.

2.16   CHROMOSOME 15

Table 2.29 summarizes 10 types of genetic diseases from chromosome 15. Table 2.30 provides the list of genes that are chosen for the natural products interactions studies. As presented in Figure 2.15a and b, various natural products impact the genes in chromosome 15. For instance, naringin inhibits the furin gene at pH 7.2 both reversibly and competitively. The authors discovered that rutin also inhibits the furin gene.

TABLE 2.29
Genetic Diseases Linked to Chromosome 15

1. Angelman syndrome
2. Prader–Willi syndrome
3. Isodicentric chromosome 15
4. Bloom syndrome
5. Breast cancer
6. Isovaleric acidemia
7. Marfan syndrome
8. Nonsyndromic deafness
9. Tay–Sachs disease
10. Tyrosinemia

TABLE 2.30
Chromosome 15 Selected Genes

B2M: Beta-2-microglobulin

This gene encodes a serum protein found in association with the major histocompatibility complex (MHC) class I heavy chain on the surface of nearly all nucleated cells. The protein has a predominantly beta-pleated sheet structure that can form amyloid fibrils in some pathological conditions. The encoded antimicrobial protein displays antibacterial activity in amniotic fluid. A mutation in this gene has been shown to result in hypercatabolic hypoproteinemia

PML: Promyelocytic leukemia

This gene is a member of the tripartite motif (TRIM) family. The TRIM motif includes three zinc-binding domains, a RING, a B-box type 1 and a B-box type 2, and a coiled-coil region. This phosphoprotein localizes to nuclear bodies where it functions as a transcription factor and tumor suppressor

FURIN: furin (paired basic amino acid cleaving enzyme)

This gene encodes a member of the subtilisin-like proprotein convertase family, which includes proteases that process protein and peptide precursors trafficking through regulated or constitutive branches of the secretory pathway. It encodes a type 1 membrane bound protease that is expressed in many tissues, including neuroendocrine, liver, gut, and brain.

IGF1R: Insulin-like growth factor 1 receptor

This receptor binds insulin-like growth factor with a high affinity. It has tyrosine kinase activity. The insulin-like growth factor I receptor plays a critical role in transformation events. Cleavage of the precursor generates alpha and beta subunits. It is highly overexpressed in most malignant tissues where it functions as an anti-apoptotic agent by enhancing cell survival. Alternatively spliced transcript variants encoding distinct isoforms have been found for this gene.

(Continued)

TABLE 2.30 (*Continued*)
Chromosome 15 Selected Genes

MAP2K1: Mitogen-activated protein kinase kinase 1

This gene is a member of the dual specificity protein kinase family, which acts as a mitogen-activated protein (MAP) kinase kinase. MAP kinases, also known as extracellular signal-regulated kinases (ERKs), act as an integration point for multiple biochemical signals. This protein kinase lies upstream of MAP kinases and stimulates the enzymatic activity of MAP kinases upon wide variety of extra- and intracellular signals. As an essential component of MAP kinase signal transduction pathway, this kinase is involved in many cellular processes such as proliferation, differentiation, transcription regulation and development

CAPN3: Calpain 3 (p94)

This gene encodes for a heterodimer consisting of a large and small subunit and is a major intracellular protease.

CHP: Calcium-Binding Protein P22

This gene encodes for a phosphoprotein that binds to the Na^+/H^+ exchanger NHE1.

FAH: Fumarylacetoacetate Hydrolase

This gene encodes for the last enzyme in the tyrosine catabolism pathway.

FBN1: Fibrillin 1

This gene encodes for a protein that is a large, extracellular matrix glycoprotein serving as a structural component of 10–12 nm calcium-binding micro-fibrils.

HEXA: Hexosaminidase A

This gene provides instructions for making one part (subunit) of an enzyme called beta-hexosaminidase A.

IVD: Isovaleryl CoA Dehydrogenase

This encodes for a mitochondrial matrix enzyme that catalyzes the third step in leucine catabolism.

CASC5: Cancer Susceptibility Candidate 5

This gene encodes a protein that is a component of the multiprotein assembly that is required for the creation of kinetochore-microtubule attachments and chromosome segregation.

OCA2: Oculocutaneous Albinism II

This gene encodes for the human homolog of the mouse p gene (pink-eye dilution).

RAD51: RAD51 Recombinase

This gene encodes for a protein that is part of the RAD51 family whose members are involved in the homologous recombination and repair of DNA.

STRC: Stereocilin

This gene encodes for a protein found in the inner ear and may be involved in hearing.

In another study, it was found that rutin combined with vitamin E to cause the down-regulation of protein expression of insulin-like growth factor receptor-1 (IGF1R), and in a different study, apigenin inhibits the activation of the IGF1R in DU145 cells. Quercetin also inhibits the IGF1R gene activity. This study suggests that quercetin could be used in treatment and prevention of hypertrophic scar and keloid.

Genistein upregulated the expression of PML and promoted the degradation of PML-RAR in NB4 cells. Furthermore, genistein significantly reversed the PML-RAR-induced misfolding of N-CoR protein. Curcumin increases the expression of the beta-2-microglobulin (B2M) gene. In another study, curcumin decreases MAP2K1 activity in diet-induced hepatic steatosis. Apigenin and genistein inhibit

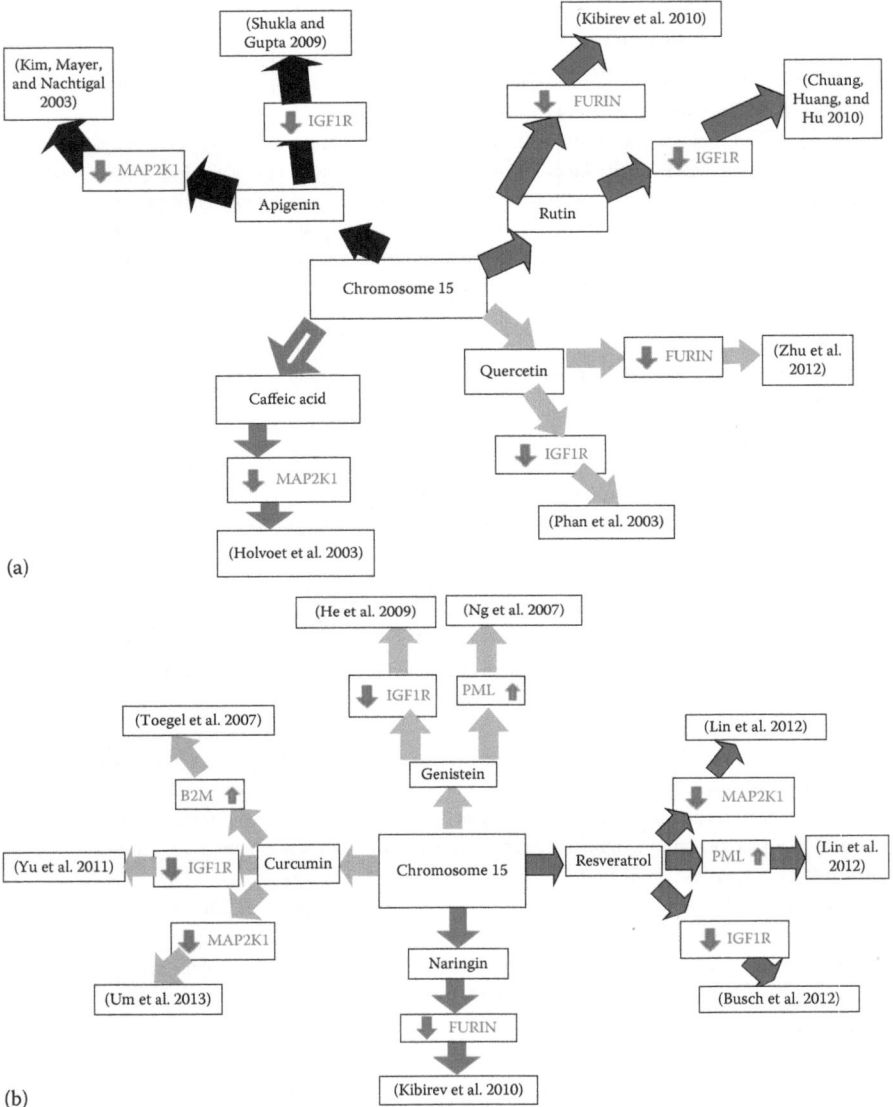

FIGURE 2.15   (See color insert.) Natural products interacting with the selected genes of chromosome 15. (a) FURIN [FURIN and quercetin]: https://www.ncbi.nlm.nih.gov/pubmed/?term=Quercetin+AND+FURIN; FURIN [FURIN and rutin]: https://www.ncbi.nlm.nih.gov/pubmed/?term=Rutin+AND+FURIN; IGF1R [IGF1R and apigenin]: https://www.ncbi.nlm.nih.gov/pubmed/?term=Apigenin+AND+IGF1R; IGF1R [IGF1R and quercetin]: https://www.ncbi.nlm.nih.gov/pubmed/?term=Quercetin++AND+IGF1R; IGF1R [IGF1R and rutin]: https://www.ncbi.nlm.nih.gov/pubmed/?term=Rutin+AND+IGF1R; MAP2K1 [MAP2K1 and apigenin]: https://www.ncbi.nlm.nih.gov/pubmed/?term=Apigenin+AND+MAP2K1; and MAP2K1 [MAP2K1 and caffeic acid]: https://www.ncbi.nlm.nih.gov/pubmed/?term= Caffeic+acid+AND+MAP2K1. (Data from Chuang, C. H. et al., *Chem Biol Interact*, 183, 434–41, 2010;   *(Continued)*

MAP2K1 in phorbol ester to stimulate macrophage differentiates of THP-1 cells. Caffeic acid inhibits MAP2K1 in human keratinocytes. Resveratrol enhances the cell death associated with mitochondrial dysfunction in bladder carcinoma cells through the inhibition of MAP2K1 gene activity. Evidently, these natural products significantly influence the genes in chromosome 15, particularly the IGF1R and MAP2K1, and the interactions of these selected natural products with other genes of chromosome 15 have not been discussed in this chapter.

2.17   CHROMOSOME 16

Table 2.31 summarizes eight types of genetic diseases from chromosome 16. Table 2.32 provides the list of genes that are chosen for the natural products interactions studies. As presented in Figure 2.16a and b, multiple natural products influence the genes in

---

TABLE 2.31
Genetic Diseases Linked to Chromosome 16

1. Trisomy 16
2. Familial Mediterranean fever (FMF)
3. Crohn's disease
4. Thalassemia
5. Autosomal dominant polycystic kidney disease (PKD-1)
6. Autism
7. Schizophrenia
8. Red hair

FIGURE 2.15 (Continued)   Holvoet, S. et al., *Exp Cell Res*, 290, 108–19, 2003; Kibirev, V. K. et al., *Ukr Biokhim Zh*, 82, 15–21, 2010; Kim, K. et al., *Biochim Biophys Acta*, 1641, 13–23, 2003; Phan, T. T. et al., *Br J Dermatol*, 148, 544–52, 2003; Shukla, S., and S. Gupta, *Mol Carcinog*, 48, 243–52, 2009; Zhu, J. et al., *Curr Med Chem*, 19, 3641–50, 2012.). (b) B2M [B2M and curcumin]: https://www.ncbi.nlm.nih.gov/pubmed/?term=Curcumin+AND+B2M; FURIN [FURIN and naringin]: https://www.ncbi.nlm.nih.gov/pubmed/?term=rutin++AND+FURIN; IGF1R [IGF1R and curcumin]: https://www.ncbi.nlm.nih.gov/pubmed/?term=Curcumin+AND+ IGF1R; IGF1R [IGF1R and genistein]: https://www.ncbi.nlm.nih.gov/pubmed/?term=Geni stein+AND+IGF1R; IGF1R [IGF1R and resveratrol]: https://www.ncbi.nlm.nih.gov/pubmed/ ?term=Resveratrol+AND+IGF1R; MAP2K1 [MAP2K1 and curcumin]: https://www.ncbi. nlm.nih.gov/pubmed/?term=Curcumin+AND+MAP2K1; MAP2K1 [MAP2K1 and resveratrol]: https://www.ncbi.nlm.nih.gov/pubmed/?term=Resveratrol+AND+MAP2K1; PML [PML and genistein]: https://www.ncbi.nlm.nih.gov/pubmed/?term=Genistein+AND+PML; and PML [PML and resveratrol]: https://www.ncbi.nlm.nih.gov/pubmed/?term=Resveratrol +AND+PML. (Data from Busch, F. et al., *J Biol Chem*, 287, 25770–81, 2012; He, H. et al., *Phytother Res*, 23, 868–73, 2009; Kibirev, V. K. et al., *Ukr Biokhim Zh*, 82, 15–21, 2010; Lin, X. et al., *Int J Urol*, 19, 757–64, 2012; Ng, A. P. et al., *Mol Cancer Ther*, 6, 2240–8, 2007; Toegel, S. et al., *BMC Mol Biol*, 8, 13, 2007; Um, M. Y., *Basic Clin Pharmacol Toxicol*, 113, 152–7, 2013; Yu, L. L. et al., *Oncol Rep*, 26, 1197–203, 2011.)

TABLE 2.32
Chromosome 16 Selected Genes

ARMC5: Armadillo Repeat-containing Protein 5

ARMC5 protein contains an armadillo repeat domain, suggesting that protein-protein interactions are important for its function. Immunohistochemical analysis of human adrenal samples, HeLa cells, and H295R adrenocortical cancer cells showed that ARMC5 localized primarily to cytoplasm.

MC1R: Melanocortin 1 Receptor

This gene encodes for a receptor that is important in normal pigmentation.

MMP-2: Matrix Metallopeptidase-2

This gene encodes for a protein part of the matrix metalloproteinase family and is involved in the breakdown of extracellular matrix in normal physiological processes.

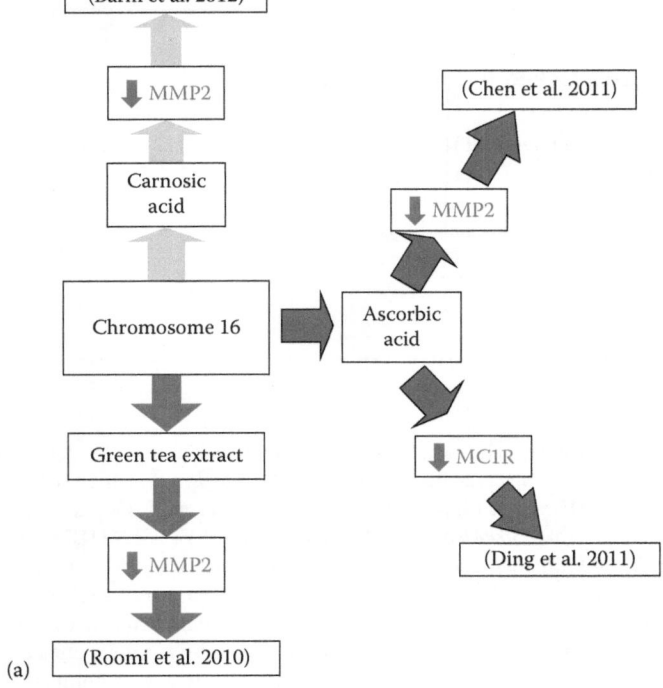

(a)

FIGURE 2.16 **(See color insert.)** Natural products interacting with the selected genes of chromosome 16. (a) MC1R [MC1R and ascorbic acid]: http://www.ncbi.nlm.nih.gov/pubmed/?term= MC1R+AND+Acid+ascorbic; MMP2 [MMP2 and ascorbic acid]: http://www.ncbi.nlm.nih. gov/pubmed/?term=MMP2+AND+Acid+ascorbic; MMP2 [MMP2 and carnosic acid]: http:// www.ncbi.nlm.nih.gov/pubmed/?term=MMP2+AND+Carnosic+acid; and MMP2 [MMP2 and green tea extract]: http://www.ncbi.nlm.nih.gov/pubmed/?term=MMP2+AND+Green+ Tea+Extract. (Data from Barni, M. V. et al., *Oncol Rep*, 27, 1041–8, 2012; Cahill, L. E., and A. El-Sohemy, *Am J Clin Nutr*, 92, 1494–500, 2010; Chen, M. F. et al., *Nutr Cancer*, 63, 1036–43, 2011; Ding, H. Y. et al., *Plant Foods Hum Nutr*, 66, 275–84, 2011; Roomi, M. W. et al., *Oncol Rep*, 24, 747–57, 2010.) *(Continued)*

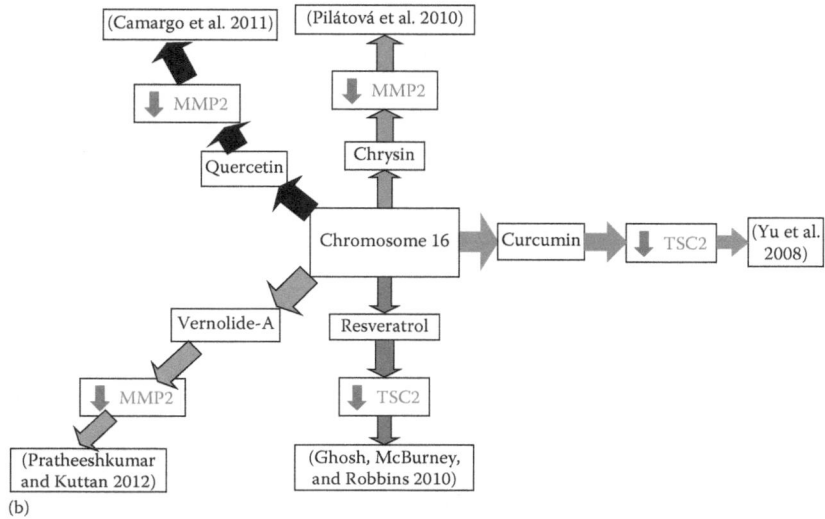

(b)

FIGURE 2.16 (Continued) Natural products interacting with the selected genes of chromosome16. (b) MMP2 [MMP2 and chrysin]: http://www.ncbi.nlm.nih.gov/pubmed/?term=MMP2+ AND+Chrysin; MMP2 [MMP2 and quercetin]: http://www.ncbi.nlm.nih.gov/pubmed/?term =MMP2+AND+Quercetin; MMP2 [MMP2 and vernolide-A]: http://www.ncbi.nlm.nih.gov/ pubmed/?term=MMP2+AND+vernolide-A; TSC2 [TSC2 and curcumin]: http://www.ncbi.nlm. nih.gov/pubmed/?term=TSC2+AND+Curcumin; and TSC2 [TSC2 and resveratrol]: http://www. ncbi.nlm.nih.gov/pubmed/?term=TSC2+AND+ Resveratrol. (Data from Camargo, C. A. et al., *Biochem Biophys Res Commun*, 406, 638–42, 2011; Ghosh, H. S. et al., *PLoS One*, 5, e9199, 2010; Pilátová, M. et al., *Gen Physiol Biophys*, 29, 134–43, 2010; Pratheeshkumar, P., and G. Kuttan, *Hum Exp Toxicol*, 31, 66–80, 2012; Yu, S. et al., *Mol Cancer Ther*, 7, 2609–20, 2008.)

chromosome 16. Several natural products interact with the matrix metallopeptidase (MMP)-2 gene involved in metastasis. For example, carnosic acid reduced MMP-2 activity in colorectal cancer cell lines. The results of this study suggest that carnosic acid could have anticancer activity. Vernolide-A, which is a sesquiterpene of lactone from *V. cinerea*, inhibited the MMP-2 protein expression in B16F-10 cells, suggesting the inhibition of metastatic progression of B16F-10 melanoma cells. Rats bearing Walker 256 carcinosarcoma were treated with 10 mg/kg quercetin, which resulted in a decrease in their MMP-2 activity, indicating that quercetin has anticancer potential. Green tea is also a potent inhibitor of MMP-2 gene. Another compound that decreases MMP-2 gene activity is ascorbic acid. This contributes to the inhibition of tumor growth and proliferation. Ascorbic acid also reduces the expression of the melanocortin 1 receptor (MC1R) gene and reduces the synthesis of melanin content in B16 cells.

Thymol exhibits no significant impact on HP gene activity in a dietary supplemental study. However, a greater interaction of ascorbic acid on HP alleles has been documented. Hp-1 allele has a greater antioxidant capacity than Hp-2 allele carriers for ascorbic acid. A Hp2-2 genotype individual has an increased risk of vitamin C deficiency if it does not meet the recommended dietary allowance for vitamin C. So far, the interaction of Hp alleles on vitamin C requirements remains unknown.

Reservertol indirectly impacts the *tuberous sclerosis* gene (TSC2) by enhancing SIRT1 activity and leads to regulate Akt/mammalian target of rapamycin (mTOR) signaling. SIRT1 is a deacetylase enzyme involved in caloric restriction-mediated longevity. In contrast, curcumin did not inhibit mTOR signaling through the TSC gene interactions rather it does inhibit through calyculin A-sensitive protein phosphatase-dependent dephosphorylation. Clearly, the mentioned natural products exhibit a significant influence on the genes in chromosome 16, particularly the MMP-2, HP, and TSC genes, and the interactions of these selected natural products with other genes of chromosome 16 have not been discussed in this chapter.

2.18 CHROMOSOME 17

Table 2.33 summarizes 38 types of genetic diseases from chromosome 17. Table 2.34 provides the list of genes that are chosen for the natural products interactions studies. As presented in Figure 2.17a and b, numerous natural products influence the

TABLE 2.33
Genetic Diseases Linked to Chromosome 17

1. 17Q21.31 microdeletion syndrome
2. Alexander disease
3. Andersen–Tawil syndrome
4. Birt–Hogg–Dubé syndrome
5. Bladder cancer
6. Breast cancer
7. Camptomelic dysplasia
8. Canavan disease
9. Cerebroretinal microangiopathy with calcifications and cysts
10. Charcot–Marie–Tooth disease
11. Charcot–Marie–Tooth disease, type 1
12. Corticobasal degeneration
13. Cystinosis
14. Depression
15. Ehlers–Danlos syndrome
16. Ehlers–Danlos syndrome, arthrochalasia type
17. Ehlers–Danlos syndrome, classical type
18. Epidermodysplasia verruciformis
19. Galactosemia
20. Glycogen storage disease, type II (Pompe disease)
21. Hereditary neuropathy with liability to pressure palsies
22. Howel–Evans syndrome
23. Li–Fraumeni syndrome
24. Maturity onset diabetes of the young, type 5
25. Miller–Dieker syndrome
26. Neurofibromatosis, type I
27. Nonsyndromic deafness

(Continued)

TABLE 2.33 (*Continued*)
Genetic Diseases Linked to Chromosome 17

28. Nonsyndromic deafness, autosomal dominant
29. Nonsyndromic deafness, autosomal recessive
30. Osteogenesis imperfecta
31. Osteogenesis imperfecta, type I
32. Osteogenesis imperfecta, type II
33. Osteogenesis imperfecta, type III
34. Osteogenesis imperfecta, type IV
35. Smith–Magenis syndrome
36. Usher syndrome
37. Usher syndrome, type I
38. Very long-chain acyl-coenzyme A dehydrogenase deficiency

TABLE 2.34
Chromosome 17 Selected Genes

GFAP: Glial fibrillary acidic protein

This gene encodes one of the major intermediate filament proteins of mature astrocytes. It is used as a marker to distinguish astrocytes from other glial cells during development. Mutations in this gene cause Alexander disease, a rare disorder of astrocytes in the central nervous system. Alternative splicing results in multiple transcript variants encoding distinct isoforms

ACTG1: Actin, Gamma 1

This gene encodes for a protein that is part of a network fibers known as actin cytoskeleton.

BRCA1: Breast Cancer 1, Early Onset

This gene encodes for a protein that helps prevent cells from growing and dividing too fast or in an uncontrolled manner.

CBX1: Chromobox Homolog 1

This gene encodes for a highly conserved non-histone protein that is part of the heterochromatin protein family.

PHB: Prohibitin

This gene is important in human cellular senescence and tumor suppression.

GALK1: Galactokinase 1

This gene encodes for an enzyme that enables the body to process galactose.

MYO15A: Myosin XVA

This gene encodes for a protein that is important for cell movement and shape.

NF1: Neurofibromin 1

This gene encodes for a protein that is produced in many cell types such as nerve cells, oligodendrocytes, and Schwann cells.

NOG: Noggin Protein

This gene encodes for a protein that is involved in the development of many body tissues such as the nerve tissue, the muscles, and the bones.

PMP22: Peripheral Myelin Protein 22

This gene encodes for a protein that is part of the connection process of the brain and spinal cord to the muscles and sensory cells.

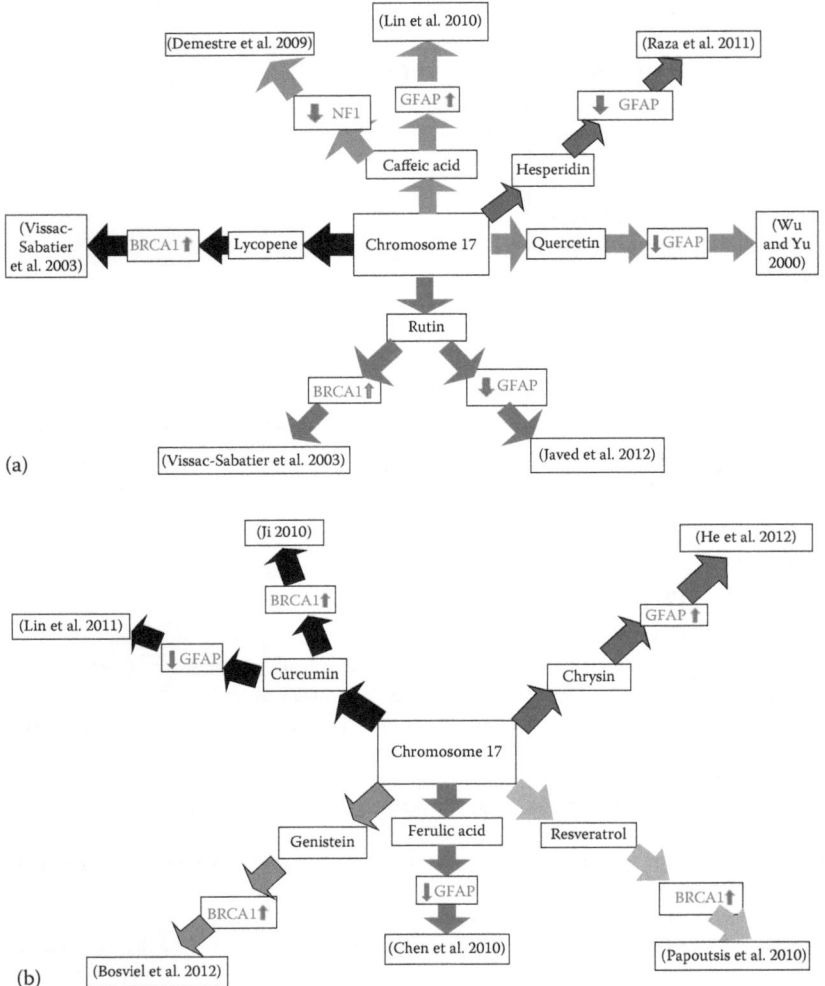

FIGURE 2.17   (See color insert.) Natural products interacting with the selected genes of chromosome 17. (a) BRCA1 [BRCA1 and lycopene]: http://www.ncbi.nlm.nih.gov/pubmed/?term= BRCA1+AND+Lycopene; BRCA1 [BRCA1 and rutin]: http://www.ncbi.nlm.nih.gov/pubmed/? term=BRCA1+AND+Rutin; GFAP; GFAP [GFAP and caffeic acid]: http://www.ncbi. nlm.nih.gov/pubmed/?term=GFAP+AND+Caffeic+acid; GFAP [GFAP and hesperidin]: http://www.ncbi.nlm.nih.gov/pubmed/?term=GFAP+AND+Hesperidin; GFAP [GFAP and quercetin]: http://www.ncbi.nlm.nih.gov/pubmed/?term=GFAP+AND+Quercetin; GFAP [GFAP and rutin]: http://www.ncbi.nlm.nih.gov/pubmed/?term=GFAP+AND+Rutin; and NF1 [NF1 and caffeic acid]: http://www.ncbi.nlm.nih.gov/pubmed/?term=NF1+AND+Caff eic+acid. (Data from Demestre, M. et al., *Phytother Res*, 23, 226–30, 2009; Javed, H. et al., *Neuroscience*, 210, 340–52, 2012; Lin, W. L. et al., *Chem Biol Interact*, 188, 607–15, 2010; Raza, S. S. et al., *Brain Res*, 1420, 93–105, 2011; Vissac-Sabatier, C. et al., *Cancer Res*, 63, 6607–12, 2003; Wu, B. Y., and A. C. Yu., *J Neurosci Res*, 62, 730–6, 2000.) (b) BRCA1 [BRCA1 and curcumin]:   *(Continued)*

genes in chromosome 17. For example, the natural product, caffeic acid, affects both neurofibromin 1 (NF1) and glial fibrillary acidic protein (GFAP) genes. The GFAP is involved in regulating the shape, movement, and function of astroglial cells that helps the oligodendocytes in myelin maintenance of nerves. Bio 30, a CAPE extract, fully suppressed the growth of a human NF1 cancer called malignant peripheral nerve sheath tumor. Another study revealed that CAPE increased the expression of GFAP gene.

Several other natural products affect the GFAP gene. Hesperidin significantly attenuated the GFAP gene. Treatment with the sodium salt of ferulic acid, sodium ferulate, combined with borneol causes a significant reduction of GFAP in C57 BL/6J mice. Chrysin significantly alleviated the proliferation of GFAP on 2VO rat models. Quercetin also affects the GFAP gene. In one study, quercetin significantly reduced the induction of GFAP in the scratch injury model. The results of this study suggest that quercetin may possess an antigliotic property. Moreover, rutin reduces the expression GFAP gene and contributes to the antigliotic property. Curcumin also impacts the GFAP gene. Curcumin downregulates the GFAP expression. This attenuates astrocyte reactivation, which could be advantageous for neuronal survival.

In addition, curcumin affects the breast cancer 1 (BRCA1) gene by increasing the expression. BRCA1 is a tumor supperessor gene. A study showed that curcumin regulates BRCA1 protein expression and modification. This indicates that curcumin has antitumor effects. Lycopene also has an effect on the BRCA1 gene. The 10-μM lycopene increases P-BRCA1 in the breast tumor cell line MCF7. Furthermore, resveratrol prevents epigenetic silencing of the BRCA1 gene by the *aryl hydrocarbon receptor* (AhR). In addition, genistein might contribute to the regulation of the BRCA1 gene. Overall, the mentioned natural products have significant impact on the genes in chromosome 17, and the interactions of these selected natural products with other genes of chromosome 17 have not been discussed in this chapter.

2.19 CHROMOSOME 18

Table 2.35 summarizes nine types of genetic diseases from chromosome 18. Table 2.36 provides the list of genes that are chosen for the natural products interactions studies. As presented in Figure 2.18, several natural products affect

FIGURE 2.17 (Continued) http://www.ncbi.nlm.nih.gov/pubmed/?term=BRCA1+AND+ Curcumin; BRCA1 [BRCA1 and genistein]: http://www.ncbi.nlm.nih.gov/pubmed/?term= BRCA1+AND+Genistein; BRCA1 [BRCA1 and resveratrol]: http://www.ncbi.nlm.nih.gov/ pubmed/?term=BRCA1+AND+Resveratrol; GFAP [GFAP and chrysin]: http://www.ncbi .nlm.nih.gov/pubmed/?term=GFAP+AND+Chrysin; GFAP [GFAP and curcumin]: http:// www.ncbi.nlm.nih.gov/pubmed/?term=GFAP+AND+Curcumin; and GFAP [GFAP and ferulic acid]: http://www.ncbi.nlm.nih.gov/pubmed/?term=GFAP+AND+Ferulic+acid. (Data from Bosviel, R. et al., *OMICS*, 16, 235–44, 2012; Chen, X. H. et al., *J Pharm Pharmacol*, 62, 915– 23, 2010; He, X. L. et al., *Eur J Pharmacol*, 680, 41–8, 2012; Ji, Z. *Breast Cancer [Auckl]*, 4, 1–3, 2010; Lin, M. S. et al., *J Surg Res*, 166, 280–9, 2011; Papoutsis, A. J. et al., *J Nutr*, 140, 1607–14, 2010.)

TABLE 2.35
Genetic Diseases Linked to Chromosome 18

1. Erythropoietic protoporphyria
2. Hereditary hemorrhagic telangiectasia
3. Niemann–Pick disease, type C
4. Porphyria
5. Selective mutism
6. Edwards syndrome (Trisomy 18)
7. Tetrasomy 18p
8. Monosomy 18p
9. Pitt–Hopkins syndrome 18q21

TABLE 2.36
Chromosome 18 Selected Genes

GRP: Gastrin-releasing peptide

This gene encodes a member of the bombesin-like family of gastrin-releasing peptides. Its preproprotein, following cleavage of a signal peptide, is further processed to produce either the 27 aa gastrin-releasing peptide or the 10 aa neuromedin C. These smaller peptides regulate numerous functions of the gastrointestinal and central nervous systems, including release of gastrointestinal hormones, smooth muscle cell contraction, and epithelial cell proliferation. These peptides are also likely to play a role in human cancers of the lung, colon, stomach, pancreas, breast, and prostate. Alternative splicing results in multiple transcript variants encoding different isoforms.

DCC: Deleted in Colorectal Cancer

This gene encodes for a netrin 1 receptor.

FECH: Ferrochelatase

This gene encodes an enzyme that is involved in producing the heme molecule.

NPC1: Niemann-Pick Disease, Type C1

This gene encodes for a protein mostly located in the membranes of the lysosomes and endosomes.

SMAD4: SMAD Family Member 4

This gene encodes for a protein involved in transmitting chemical signals from the cell surface to the nucleus.

KC6: Keratoconus Gene 6

This gene is an RAN gene and is associated with keratoconus and obesity.

the genes in chromosome 18. For instance, apigenin reduces the expression of the gastrin-releasing peptide (GRP) gene. Resveratrol also has an impact on the GRP gene. Resveratrol induced the upregulation of GRP. This indicates that resveratrol induces endoplasmic reticulum (ER) stress that might lead to cell death. Quercetin affects the myelin basic protein (MBP) gene. This MBP is the constituent of the myelin sheath of oligodendrocytes and Schwann cells of the nervous system. Oral administration of 10-mg/Kg quercetin induces a progressive decrease in MBP, which may have an effect on the nervous system. Caffeic

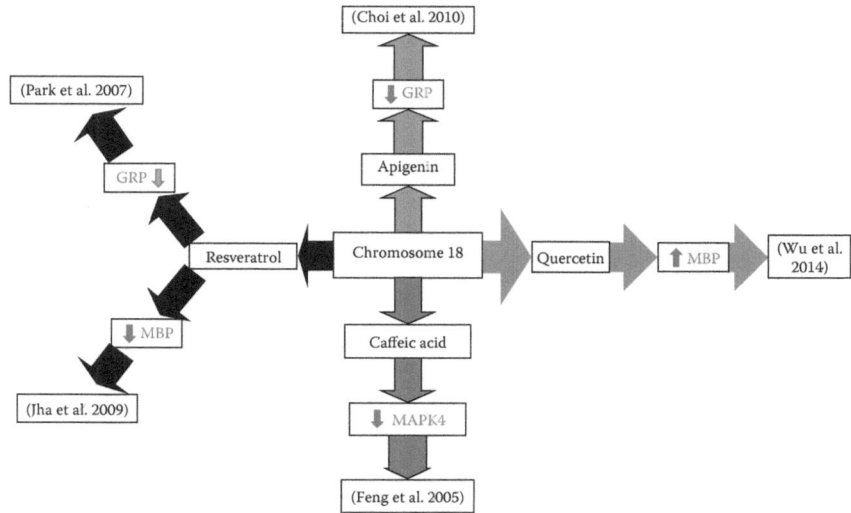

FIGURE 2.18 **(See color insert.)** Natural products interacting with the selected genes of chromosome 18. GRP [GRP and apigenin]: http://www.ncbi.nlm.nih.gov/pubmed/?term=GRP+AND+Apigenin; MAPK4 [MAPK4 and caffeic acid]: http://www.ncbi.nlm.nih.gov/pubmed/?term=MAPK4+AND+Caffeic+acid; MBP [MBP and quercetin]: http://www.ncbi.nlm.nih.gov/pubmed/24519463; MBP [MBP and resveratol]: http://www.ncbi.nlm.nih.gov/pubmed/?term=19696693; and GRP [GRP and resveratol]: http://www.ncbi.nlm.nih.gov/pubmed/?term=GRP+AND+Resveratrol. (Data from Choi, A. Y. et al., *Neurochem Int*, 57, 143–52, 2010; Feng, R. et al., *J Biol Chem*, 280, 27888–95, 2005; Jha, R. K. et al., *Pancreas*, 38, 947–53, 2009; Park, J. W. et al., *Oncol Rep*, 18, 1269–73, 2007; Wu, X. et al., *Cell Mol Neurobiol*, 34, 463–71, 2014.)

acid has an effect on the MAPK4 gene. One study suggests that chlorogenic acid, the ester of caffeic acid with quinic acid, may have chemoprevenative effects through the suppression of MAPK activation. Clearly, these natural products have a significant influence on genes in chromosome 18, and the interactions of these selected natural products with other genes of chromosome 18 have not been discussed in this chapter.

2.20 CHROMOSOME 19

Table 2.37 summarizes 20 types of genetic diseases from chromosome 19. Table 2.38 provides the list of genes that are chosen for the natural products interactions studies. As presented in Figure 2.19, several natural products affect the genes in chromosome 19. Carnosic acid affects the PRX gene that codes for periaxin protein, which is required for maintenance of peripheral nerve myelin sheath. Carnosic acid reduces the formation of PRX. Futhermore, ascorbic acid reduces the PRX gene. Quercetin also affects the PRX gene. Dr. Chaudhary and colleagues found

TABLE 2.37
Genetic Diseases Linked to Chromosome 19

1. Alzheimer's disease
2. CADASIL
3. Centronuclear myopathy autosomal dominant form
4. Charcot–Marie–Tooth disease
5. Congenital hypothyroidism
6. Familial hemiplegic migraine
7. Glutaric acidemia, type 1
8. Hemochromatosis
9. Leber's congenital amaurosis
10. Maple syrup urine disease
11. Multiple epiphyseal dysplasia
12. Myotonic dystrophy
13. Myotubular myopathy autosomal dominant form
14. Marfan syndrome
15. Oligodendroglioma
16. Peutz–Jeghers syndrome
17. Pseudoachondroplasia
18. Spinocerebellar ataxia, type 6
19. X-linked agammaglobulinemia or Bruton's disease
20. Prolidase deficiency

TABLE 2.38
Chromosome 19 Selected Genes

APOE: Apolipoprotein E
   This gene encodes for a protein that combines with fats in the body to produce lipoproteins.
BCKDHA: Branched Chain Keto Acid Dehydrogenase E1, Alpha Polypeptide
   This gene helps in making the alpha subunit of an enzyme complex.
COMP: Cartilage Oligomeric Matrix Protein
   This gene encodes for the protein found in the extracellular matrix, which is a complex lattice of proteins and other molecules that forms in the space between cells.
DMPK: Dystrophia Myotonica-Protein Kinase
   This gene encodes for a protein that may be involved in muscle, heart, and brain cells.
GCDH: Glutaryl-Coenzyme A Dehydrogenase
   This gene encodes for an enzyme found in the mitochondria and is involved in the breakdown of the amino acids lysine, hydroxylysine, and tryptophan.
HAMP: Hepcidin Antimicrobial Peptide
   This gene encodes for a protein that is important for iron balance in the body.
FKRP: Fukutin-Related Protein
   This gene encodes for a protein that is part of many tissues of the body and is mostly present in the muscles for movement, the brain, and the heart.

(Continued)

TABLE 2.38 (*Continued*)

Chromosome 19 Selected Genes

NRTN: Neurturin

This gene signals through RET and a GPI-linked co-receptor and promotes survival of neuronal populations.

PRX: Periaxin

This gene encodes for a protein that is important for the maintenance of myelin.

STK11: Serine/Threonine Kinase 11

This gene encodes for an enzyme that is a tumor suppressor.

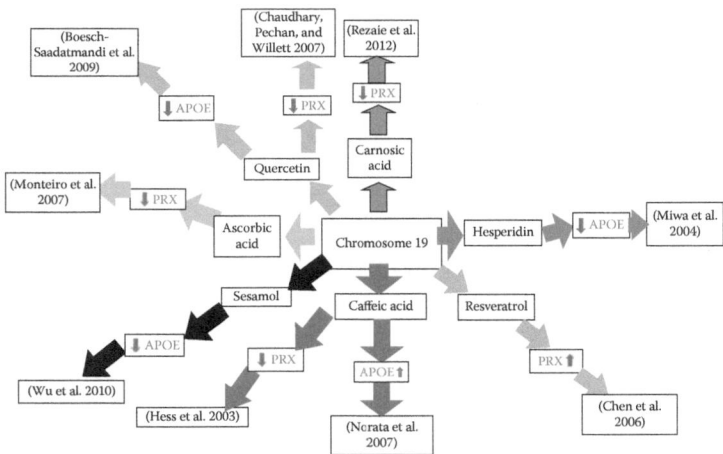

FIGURE 2.19 (**See color insert.**) Natural products interacting with the selected genes of chromosome 19. APOE [APOE and caffeic acid]: http://www.ncbi.nlm.nih.gov/pubmed/?term=APOE+AND+Caffeic+acid; APOE [APOE and hesperidin]: http://www.ncbi.nlm.nih.gov/pubmed/?term=APOE+AND+Hesperidin; APOE [APOE and quercetin]: http://www.ncbi.nlm.nih.gov/pubmed/?term=APOE+AND+Quercetin; APOE [APOE and sesamol]: http://www.ncbi.nlm.nih.gov/pubmed/?term=APOE+AND+Sesamol; PRX [PRX and ascorbic acid]: http://www.ncbi.nlm.nih.gov/pubmed/?term=PRX+AND+Acid+; PRX [PRX and caffeic acid]: http://www.ncbi.nlm.nih.gov/pubmed/?term=PRX+AND+Caffeic+acid; PRX [PRX and carnosic acid]: http://www.ncbi.nlm.nih.gov/pubmed/?term=PRX+AND+ Carnosic+acid; and PRX [PRX and resveratrol]: http://www.ncbi.nlm.nih.gov/pubmed/?term=PRX+AND+Resveratrol. (Data from Boesch-Saadatmandi, C. et al., *Br J Nutr*, 101, 1440–3, 2009; Chaudhary, A. et al., *Toxicol Appl Pharmacol*, 220, 197–210, 2007; Chen, H. B. et al., *J Cell Biochem*, 97, 314–26, 2006; Hess, A. et al., *J Biol Chem*, 278, 45419–34, 2003; Horio, F. et al., *J Nutr Sci Vitaminol [Tokyo]*, 52, 28–32, 2006; Kwon, E. Y. et al., *J Med Food*, 12, 996–1003, 2009; Miwa, Y. et al., *J Nutr Sci Vitaminol [Tokyo]*, 50, 211–18, 2004; Monteiro, G. et al., *Proc Natl Acad Sci U S A*, 104, 4886–91, 2007; Mulik, R. S. et al., *Int J Pharm*, 437, 29–41, 2012; Norata, G. D., *Atherosclerosis*, 191, 265–71, 2007; Rezaie, T. et al., *Invest Ophthalmol Vis Sci*, 53, 7847–54, 2012; Shin, S. M. et al., *Mol Pharmacol*, 76, 884–95, 2009; Wu, W. H. et al., *Mol Nutr Food Res*, 54, 1340–50, 2010.)

that quercetin inhibited both the BaP-mediated effects on PRX and BaP-mediated upregulation of PRX I. Moreover, the quercetin also neutralized BaP-mediated downregulation of PRX II.

The natural products caffeic acid, quercetin, and hesperidin impact the apolipoprotein E (APOE) gene. Caffeic acid significantly decreases atherosclerosis in APOE KO mice, indicating that caffeic acid is APOE-like activity. In addition, dietary quercetin significantly lowered levels of TNF-alpha in APOE3 mice. Hesperidin decreases the levels of APOE in hyperlipidemic subjects. Overall, these natural products have considerable impact on the genes in chromosome 19, and the interactions of these selected natural products with other genes of chromosome 19 have not been discussed in this chapter.

2.21   CHROMOSOME 20

Table 2.39 summarizes 11 types of genetic diseases from chromosome 20. Table 2.40 provides the list of genes that are chosen for the natural products interactions studies. As presented in Figure 2.20, several natural products affect the genes in chromosome 20.

TABLE 2.39
Genetic Diseases Linked to Chromosome 20

1. Albright's hereditary osteodystrophy
2. Arterial tortuosity syndrome
3. Adenosine deaminase deficiency
4. Alagille syndrome
5. Celiac disease
6. Galactosialidosis—CTSA
7. Maturity onset diabetes of the young, type 1
8. Neuronal ceroid lipofuscinosis
9. Pantothenate kinase-associated neurodegeneration
10. Transmissible spongiform encephalopathy (prion diseases)
11. Waardenburg syndrome

TABLE 2.40
Chromosome 20 Selected Genes

AHCY: Adenosylhomocysteine
   This gene encodes for the enzyme S-adenosylhomocystein hydrolase.
BMP2: Bone Morphogenetic Protein 2
   This gene encodes for a protein that is part of the transforming growth factor, beta superfamily.

(Continued)

TABLE 2.40 (*Continued*)
Chromosome 20 Selected Genes

DNAJC5: Cysteine String Protein

This gene encodes for a protein that is found in the brain and is important for the transmission of nerve impulses.

EDN3: Endothelin 3

This gene encodes for a protein that is produced in various cells and tissues and is involved in the development and function of blood vessels, the production of some hormones, and the stimulation of cell growth and division.

GSS: Glutathione Synthetase

This gene encodes for an enzyme that participated in the gamma-glutamyl cycle.

JAG1: Jagged 1

This gene encodes for the protein involved in an important pathway in which cells can signal each other.

PANK2: Pantothenate Kinase 2

This gene encodes for an enzyme that is active in mitochondria.

PRNP: Prion Protein

This gene encodes for a protein that is active in the brain and many other tissues.

SALL4: Sal-Like 4 (Drosophila)

This gene encodes for a protein that might be a zinc-finger transcription factor.

VAMP: (Vesicle-Associate Membrane Protein)—Associated Protein B and C

This gene encodes for a type IV membrane protein that is found in the plasma and intracellular vesicle membranes.

For instance, the natural product resveratrol impacts the prion protein (PRNP) gene. Resveratrol reversed mutant PRNP neurotoxicity. Furthermore, resveratrol reversed cell death instigated by mutant PRNP in cerebellar granule neurons from PRNP-null mice.

Multiple other natural products impact the bone morphogenetic protein 2 (BMP-2) gene. BMP-2 plays a key role in skeletal development, repair, and regeneration. Naringin increases the expression of the BMP-2 gene in osteoblasts. Resveratrol also induces BMP-2 gene expression in both in vivo and in vitro models. Moreover, the curcumin analog, UBS109, stimulates BMP-2 activity in an in vitro tissue culture model. Hesperidin also affects the BMP-2 gene. Hesperidin significantly induced mRNA expression of the BMP-2 genes after 48 hours of exposure. Quercetin, apigenin, and genistein also increase the BMP-2 gene activity in in vitro models. Ascorbic acid does not interact with BMP-2 gene product in human adipose-derived stem cells. Overall, the mentioned natural products exhibit a significant impact on the genes in chromosome 20, particularly the BMP-2 gene, and the interactions of these selected natural products with other genes of chromosome 20 have not been discussed in this chapter.

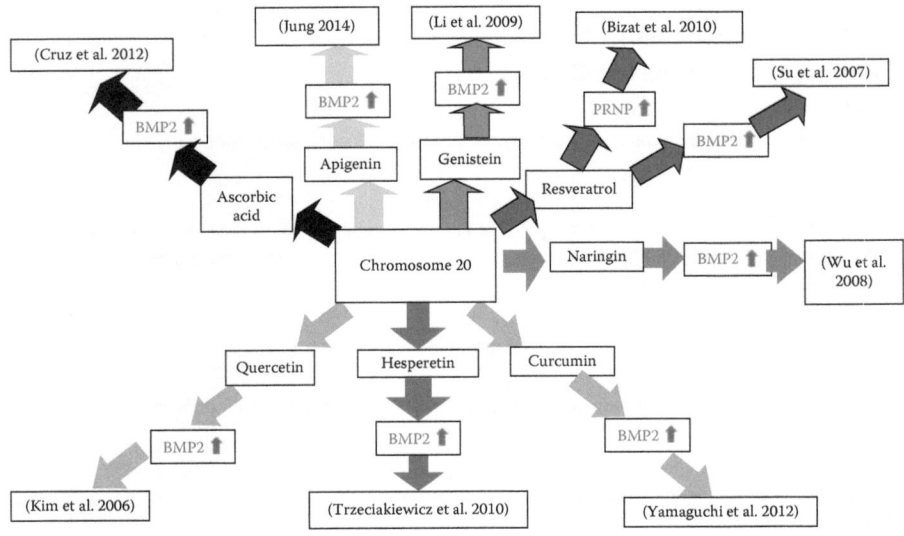

FIGURE 2.20 (See color insert.) Natural products interacting with the selected genes of chromosome 20. BMP2 [BMP2 and ascorbic acid]: http://www.ncbi.nlm.nih.gov/ pubmed/?term=BMP2+AND+Acid+ascorbic; BMP2 [BMP2 and apigenin]: http:// www.ncbi.nlm.nih.gov/pubmed/?term=BMP2+AND+Apigenin; BMP2 [BMP2 and curcumin]: http://www.ncbi.nlm.nih.gov/pubmed/?term=BMP2+AND+Curcumin; BMP2 [BMP2 and genistein]: http://www.ncbi.nlm.nih.gov/pubmed/?term=BMP2+AND+ Genistein; BMP2 [BMP2 and hesperetin]: http://www.ncbi.nlm.nih.gov/pubmed/?term =BMP2+AND+Hesperetin; BMP2 [BMP2 and naringin]: http://www.ncbi.nlm.nih.gov/ pubmed/?term=BMP2+AND+Naringin; BMP2 [BMP2 and quercetin]: http://www. ncbi.nlm.nih.gov/pubmed/?term=BMP2+AND+Quercetin; BMP2 [BMP2 and resveratrol]: http://www.ncbi.nlm.nih.gov/pubmed/?term=BMP2+AND+Resveratrol; and PRNP [PRNP and resveratrol]: http://www.ncbi.nlm.nih.gov/pubmed/?term=PRNP+AND+Resv eratrol. (Data from Bizat, N. et al., *J Neurosci*, 30, 5394–403, 2010; Cruz, A. C. et al., *J Appl Oral Sci*, 20, 628–35, 2012; Jung, W. W., *Int J Mol Med*, 33, 1327–34, 2014; Kim, Y. J. et al., *Biochem Pharmacol*, 72, 1268–78, 2006; Li, X. et al., *J Biomol Screen*, 14, 1251–6, 2009; Su, J. L. et al., *J Biol Chem*, 282, 19385–98, 2007; Trzeciakiewicz, A. et al., *J Nutr Biochem*, 21, 424–31, 2010; Wu, J. B. et al., *Eur J Pharmacol*, 588, 333–41, 2008; Yamaguchi, M. et al., *Integr Biol [Camb]*, 4, 905–13, 2012; Yang, Y. et al., *J Inherit Metab Dis*, 28, 1055–64, 2005.)

2.22 CHROMOSOME 21

Table 2.41 summarizes 11 types of genetic diseases from chromosome 21. Table 2.42 provides the list of genes that are chosen for the natural products interactions studies. As presented in Figure 2.21, multiple natural products impact the genes in chromosome 21. For example, many natural products affect the superoxide dismutase 1 (SOD1) gene. One natural product that interacts with the SOD1 gene is caffeic acid. In a study, mice expressing a mutant superoxide dismutase (SOD1 G93A) linked

TABLE 2.41
Genetic Diseases Linked to Chromosome 21

1. Alzheimer's disease
    a. Alzheimer's disease, type 1
2. Amyotrophic lateral sclerosis
    a. Amyotrophic lateral sclerosis, type 1
3. Down's syndrome
4. Erondu–Cymet syndrome
5. Holocarboxylase synthetase deficiency
6. Homocystinuria
7. Jervell and Lange-Nielsen syndrome
8. Leukocyte adhesion deficiency
9. Majewski osteodysplastic primordial dwarfism, type II (MOPD II or MOPD2)
10. Nonsyndromic deafness
    a. Nonsyndromic deafness, autosomal recessive
11. Romano–Ward syndrome

TABLE 2.42
Chromosome 21 Selected Genes

AIRE: Autoimmune Regulator

This gene encodes for a protein that is active in the thymus.

APP: Amyloid Beta (A4) Precursor Protein

This gene encodes for a cell surface receptor and transmembrane precursor protein that is cleaved by secretases to form a number of peptides.

C21orf59: Chromosome 21 Open Reading Frame 59

This gene encodes for a protein that is important in the assembly of dynein arm and motile cilia function.

CBS: Cystathionine-Beta-Synthase

This gene encodes for a protein that acts as a homotetramer to catalyze the conversion of homocysteine to cystathionine, which is the first step in the transsulfuration pathway.

CLDN14: Claudin 14

This gene encodes for a protein that provides building components for tight junctions.

HLCS: Holocarboxylase Synthetase

This gene encodes for an enzyme that is important for the effective use of biotin.

KCNE1: Potassium Voltage-Gated Channel, ISK-Related Family, Member 1

This gene encodes for a protein that regulates the activity of potassium channels.

SOD1: Superoxide Dismutase 1

This gene encodes for a protein that binds copper and zinc ions and is one of two isozymes responsible for destroying free superoxide radicals in the body.

TMPRSS3: Transmembrane Protease, Serine 3

This gene encodes for a protein that is part of the serine protease family.

PCNT: Centrosomal Pericentrin

This gene encodes for a protein that binds to calmodulin and is expressed in the centrosome.

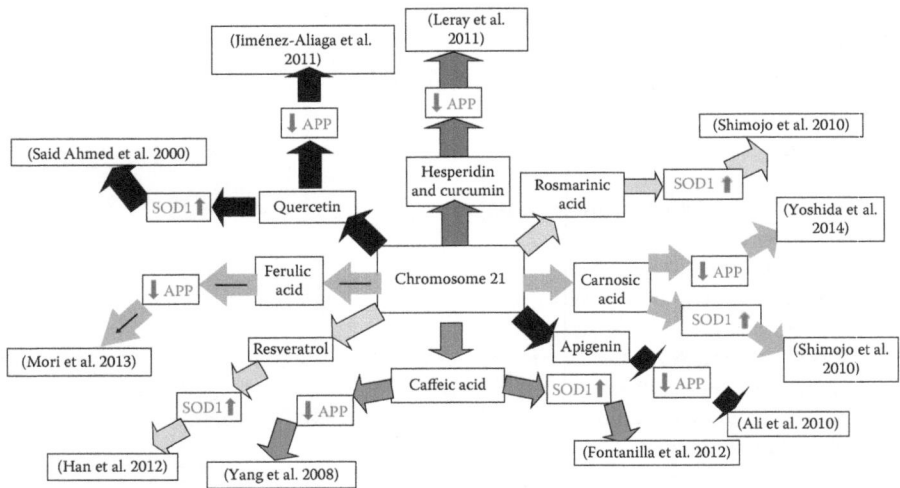

FIGURE 2.21   (See color insert.) Natural products interacting with the selected genes of chromosome 21. APP [APP and apigenin]: http://www.ncbi.nlm.nih.gov/pubmed/?term=APP+ AND+Apigenin; APP [APP and caffeic acid]: http://www.ncbi.nlm.nih.gov/pubmed/?term= APP+AND+Caffeic+acid; APP [APP and carnosic acid]: http://www.ncbi.nlm.nih.gov/pubmed/? term=APP+AND+Hesperidin+%26Curcumin; APP [APP and ferulic acid]: http://www.ncbi. nlm.nih.gov/pubmed/?term=APP+AND+Ferulic+acid; APP [APP and hesperidin & curcumin]: http://www.ncbi.nlm.nih.gov/pubmed/?term=APP+AND+Hesperidin+%26Curcumin; APP [APP and quercetin]: http://www.ncbi.nlm.nih.gov/pubmed/?term=APP+AND+Quercetin; SOD1 [SOD1 and caffeic acid]: http://www.ncbi.nlm.nih.gov/pubmed/?term=SOD1+AND+ Caffeic+acid; SOD1 [SOD1 and carnosic acid]: http://www.ncbi.nlm.nih.gov/pubmed/?term= APP+AND+Hesperidin+%26Curcumin; SOD1 [SOD1 and quercetin]: http://www.ncbi.nlm. nih.gov/pubmed/?term=SOD1+AND+Quercetin; SOD1 [SOD1 and resveratrol]: http://www. ncbi.nlm.nih.gov/pubmed/?term=SOD1+AND+Resveratrol; SOD1 [SOD1 and rosmarinic acid]: http://www.ncbi.nlm.nih.gov/pubmed/?term=APP+AND+Hesperidin+%26Curcumin.   (Data from Ali, S. et al., *Int J Biochem Cell Biol*, 42, 113–19, 2010; Fontanilla, C. V. et al., *Neuroscience*, 205, 185–93, 2012; Han, S. et al., *Brain Res*, 1483, 112–17, 2012; Jiménez-Aliaga, K. et al., *Life Sci*, 89, 939–45, 2011; Leray, V. et al., *Br J Nutr*, 106, S198–201, 2011; Mori, T. et al., *PLoS One*, 8, e55774, 2013; Noll, C. et al., *J Nutr Biochem*, 20, 586–96, 2009; Said Ahmed, M. et al., *J Neurol Sci*, 176, 88–94, 2000; Shimojo, Y. et al., *J Neurosci Res*, 88, 896–904, 2010; Yang, J. Q. et al., *CNS Neurosci Ther*, 14, 10–16, 2008; Yoshida, H. et al., *Neurosci Res*, 79, 83–93, 2014.)

to human amyotrophic lateral sclerosis were treated with caffeic acid. These mice exhibited lower levels of phosphorylated p38, which is involved in both inflammation and neuronal death. Rosmarinic acid has preventive effects for amyotrophic lateral sclerosis in the human SOD1 G93A transgenic mouse model. Resveratrol could protect motor neurons from the mutant SOD1-induced neurotoxicity in transgenic mice overexpressing SOD1 G93A. Carnosic acid also affects the SOD1 gene. In one study, rosemary extract, which contains carnosic acid, was administered to human

SOD1 G93A transgenic mice ALS model. The carnosic acid significantly delayed motor dysfunction in paw grip endurance tests, attenuated the degeneration of motor neurons, and extended the life span of these mice.

Various natural products also affect the amyloid beta (A4) precursor protein (APP) gene. Carnosic acid enhanced the knockdown of TACE (ADAM gene that is coding for disintegrin and metalloprotease domain) by siRNA-reduced soluble-APPα release. In another study, caffeic acid was administered intragastrically at 30 minutes prior to microinjection of aluminum in mice. The caffeic acid prevented neuronal death in APP caused by aluminum overload. Furthermore, curcumin and hesperidin also inhibit the APP gene. A different study shows that apigenin affects the NF-kappaB in the IL1-induced expression of C/EBP delta, which regulates the downstream APP gene.

Quercetin also affects the APP gene. Quercetin almost completely decreased ROS generation in these APPswe cells. Well-characterized mutant human APP-overexpressing murine neuron-like cells are studied with ferulic acid. The authors discovered that the ferulic acid reduced amyloidogenic APP proteolysis in these cells. Thus, numerous natural products affect genes in chromosome 21, chiefly the APP and SOD1 genes, and the interactions of these selected natural products with other genes of chromosome 21 have not been discussed in this chapter.

2.23　CHROMOSOME 22

Table 2.43 summarizes 13 types of genetic diseases from chromosome 22. Table 2.44 provides the list of genes that are chosen for the natural products interactions studies. As presented in Figure 2.22a and b, various natural products affect the genes in

TABLE 2.43
Genetic Diseases Linked to Chromosome 22

1. Amyotrophic lateral sclerosis
2. Breast cancer
3. DiGeorge syndrome
4. Desmoplastic small round cell tumor
5. 22q11.2 deletion syndrome
6. 22q13 deletion syndrome or Phelan–McDermid syndrome
7. Li–Fraumeni syndrome
8. Neurofibromatosis, type 2
9. Rubinstein–Taybi syndrome
10. Waardenburg syndrome
11. Cat eye syndrome
12. Methemoglobinemia
13. Schizophrenia

TABLE 2.44
Chromosome 22 Selected Genes

ARSA: Arylsulfatase A
   This gene encodes for an enzyme that is located in lysosomes.
TBX1: T-Box 1
   This gene is important in the formation of tissues and organs during embryonic development.
LARGE: Like-Glycosyltransferase
   This gene encodes for a protein involved in glycosylation.
PDGFB: Platelet-Derived Growth Factor Beta Polypeptide
   This gene encodes for an isoform of the platelet-derived growth factor protein.
NEFH: Neurofilament, Heavy Polypeptide
   This gene encodes for a protein that is part of a structural framework helping in defining the shape
   and size of neurons.
CHEK2: Checkpoint Kinase 2
   This gene encodes for a protein that acts as a tumor suppressor.
NF2: Neurofibromin 2 (Merlin)
   This gene encodes for a protein made in the nervous system, particularly in Schwann cells.
SOX10: SRY (Sex-Determining Region Y), Box 10
   This gene is important for the formation of tissues and organs during embryonic development.
EP300: E1A-Binding Protein p300
   This gene encodes for a protein that regulates the activity of many genes in tissues throughout the
   body.
TYMP: Thymidine Phosphorylase
   This gene encodes for a protein that is used as a building block of DNA.

chromosome 22. Quercetin inhibits catechol-O-methyltransferase (COMT) activity properties that might potentiate the anticatatonic effect of L-dopa plus carbidopa treatment. COMT is involved in the breakdown of neurotransmitters that conduct signals from one nerve cell to another. This suggests that quercetin may serve as an effective adjunct to L-dopa therapy in Parkinson's disease. A different study showed that diosmin inhibits COMT activity in the mesenteric vein. Green tea extract increases COMT activity in human clinical trial.

The natural product resveratrol impacts the E1A-binding protein p300 (EP300) gene and the COMT gene. Resveratrol downregulates the p300 protein in dystrophin-deficient mice and improves the cardiomyopathy. Another study exhibited that resveratrol increased uterine artery blood flow velocity and fetal weight in COMT$^{-/-}$ mice. It was concluded from this study that resveratrol shows potential as a therapeutic treatment for pre-eclampsia and fetal growth restriction.

Genistein also affects the COMT genes. The combination of genistein and diindolylmethane (DIM) altered major E2 metabolism pathways in LNCaP and PC3 (E2-insensitive) prostate cancer cells by amplifying the expression of COMT. This study suggests that DIM and genistein decrease the effects of E2, which have the potential to promote prostate cancer.

Ascorbic acid affects the COMT and SOX10 genes (the transcription factor and a key regulator in differentiation of peripheral glial cells). Ascorbic acid serves as a weak competitive inhibitor of COMT in Parkinson's disease. Chrysin, genistein, and flavone have no increase in COMT activity in the cytosol fraction of healthy human mammary tissues.

Curcumin affects the EP300 and neurofibromin 2 (NF2) gene. Curcumin inhibits the activity and expression of p300 in B-NHL cells. A study demonstrated that the combination of curcumin and a heat shock protein inhibitor downregulates the proliferation of a human schwannoma cell line harboring NF2 mutation. Overall, a plethora of natural products affect various genes in chromosome 22, and the interactions of these selected natural products with other genes of chromosome 22 have not been discussed in this chapter.

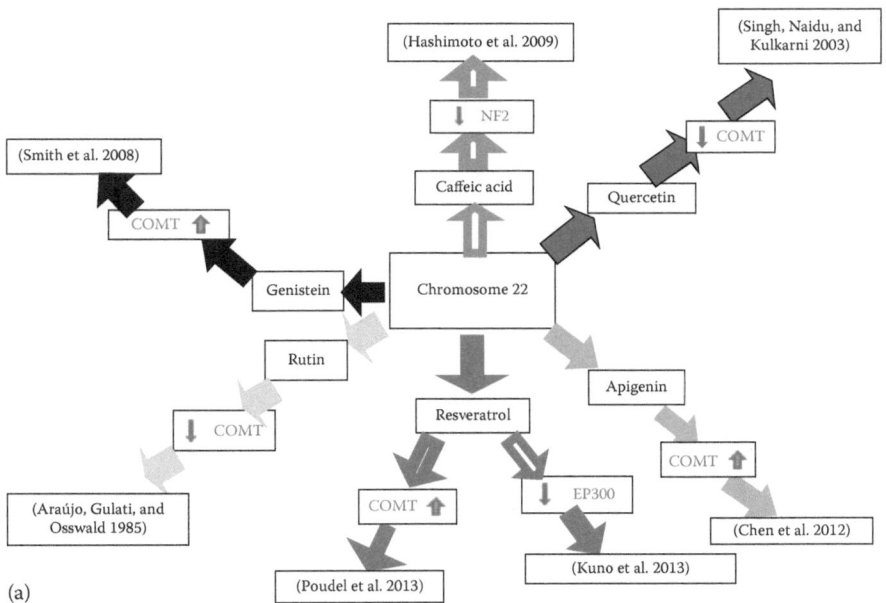

(a)

FIGURE 2.22 (See color insert.) Natural products interacting with the selected genes of chromosome 22. (a) NF2 [NF2 and caffeic acid]: http://www.ncbi.nlm.nih.gov/pubmed/?term=NF2+AND+Caffeic+acid; COMT [COMT and quercetin]: http://www.ncbi.nlm.nih.gov/pubmed/?term=COMT+AND+Quercetin; COMT [COMT and apigenin]: http://www.ncbi.nlm.nih.gov/pubmed/?term=COMT+AND+Apigenin; EP300 [EP300 and resveratrol]: http://www.ncbi.nlm.nih.gov/pubmed/?term=EP300+AND+Resveratrol; COMT [COMT and resveratrol]: http://www.ncbi.nlm.nih.gov/pubmed/23667712; COMT [COMT and rutin]: http://www.ncbi.nlm.nih.gov/pubmed/?term=COMT+AND+Rutin; and COMT [COMT and genistein]: http://www.ncbi.nlm.nih.gov/pubmed/?term=COMT+AND+Genistein. (Data from Araújo, D. et al., *Arch Int Pharmacodyn Ther*, 277, 192–202, 1985; Chen, Z. et al., *Fitoterapia*, 83, 1616–22, 2012; Hashimoto, H. et al., *Drug Discov Ther*, 3, 243–6, 2009; Kuno, A. et al., *J Biol Chem*, 288, 5963–72, 2013; Poudel, R. et al., *PLoS One*, 8, e64401, 2013; Singh, A. et al., *Pharmacology*, 68, 81–8, 2003; Smith, S. et al., *J Nutr*, 138, 2379–85, 2008.) *(Continued)*

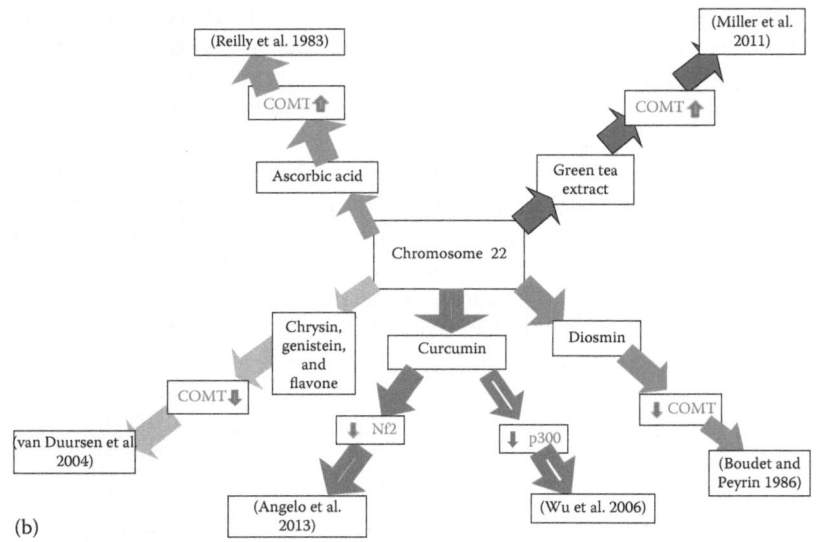

FIGURE 2.22 (Continued) Natural products interacting with the selected genes of chromosome 22. (b) COMT [COMT and ascorbic acid]: http://www.ncbi.nlm.nih.gov/pubmed/?term=COMT +AND+Acid+ascorbic; COMT [COMT and chrysin, genistein, and flavones]: http://www.ncbi. nlm.nih.gov/pubmed/?term=COMT+AND+Chrysin%2C+Genistein%2C+and+flavone.; COMT [COMT and diosmin]: http://www.ncbi.nlm.nih.gov/pubmed/?term=COMT+AND+Diosmin; COMT [COMT and green tea extract]: http://www.ncbi.nlm.nih.gov/pubmed/?term= COMT+AND+Green+tea+extract; Nf2 [Nf2 and curcumin]: http://www.ncbi.nlm.nih.gov/ pubmed/?term=Nf2+AND+Curcumin; and p300 [p300 and curcumin]: http://www.ncbi.nlm. nih.gov/pubmed/?term=p300+AND+Curcumin. (Data from Angelo, L. S. et al., *Bioorg Med Chem*, 21, 932–9, 2013; Boudet, C., and L. Peyrin, *Arch Int Pharmacodyn Ther*, 283, 312–20, 1986; Miller, R. J. et al., *Br J Nutr*, 105, 1138–44, 2011; Reilly, D. K. et al., *Adv Neurol*, 37, 51–60, 1983; van Duursen, M. B., *Toxicol Sci*, 81, 316–24, 2004; Wu, Q. et al., *Zhongguo Shi Yan Xue Ye Xue Za Zhi*, 14, 293–7, 2006.)

2.24　CHROMOSOME X

Table 2.45 summarizes three types of genetic diseases from chromosome X. Table 2.46 provides the list of genes that are chosen for the natural products interactions studies. As presented in Figure 2.23a and b, many natural products affect the genes in chromosome X. For instance, genistein induces the Bruton agammaglobulinemia tyrosine kinase (BTK) gene, which is involved in G2/M arrest and apoptosis of various cancer cells. Various natural products impact the emopamil-binding protein (sterol isomerase) (EBP) gene. Curcumin increased the expression of EBP in human CD4+ and Jurkat T cells. A different study showed that phyllanthus amarus inhibits the enhancer-binding protein (C/EBP alpha and beta) and mediated the upregulation of HBV enhancer I activity in a dose-dependent manner. In addition, genistein inhibits C/EBP beta activity by increasing the level of C/EBP homologous protein and probably by inhibiting the tyrosine phosphorylation of C/EBP beta. Resveratrol reduced C/EBP beta mRNA levels in 3T3-L1 adipocytes. In addition,

TABLE 2.45
Genetic Diseases Linked to Chromosome X

1. Klinefelter's syndrome
2. Triple X syndrome (also called 47,XXX or trisomy X)
3. Turner's syndrome

TABLE 2.46
Chromosome X Selected Genes

DMD: Dystrophin

The dystrophin gene is the largest gene found in nature, measuring 2.4 Mb. The gene was identified through a positional cloning approach, targeted at the isolation of the gene responsible for Duchenne (DMD) and Becker (BMD) Muscular Dystrophies. DMD is a recessive, fatal, X-linked disorder occurring at a frequency of about 1 in 3,500 new-born males

EBP: Emopamil binding protein

This gene is an integral membrane protein of the endoplasmic reticulum. It is a high affinity binding protein for the antiischemic phenylalkylamine Ca2+ antagonist [3H]emopamil and the photoaffinity label [3H]azidopamil

BTK: Bruton Agammaglobulinemia Tyrosine Kinase

This gene encodes for a protein that is essential for the development and maturation of B cells.

EDA: Ectodysplasin A

This gene encodes for a protein that is part of a signaling pathway, which is important for the development before birth.

F8: Coagualtion Factor VIII, Procoagulant Component

This gene encodes for a protein that is essential for the formation of blood clots.

MTM1: Myotubularin 1

This gene encodes for a protein involved in the development and maintenance of muscle cells.

PHKA2: Phosphorylase Kinase, Alpha 2 (Liver)

This gene helps in the production of the alpha subunit of the phosphorylase B kinase enzyme.

TAZ: Tafazzin

This gene encodes for a protein that is important for the maintenance of the inner membrane of mitochondria and promotes the differentiation and maturation of osteoblasts and prevents adipocytes from maturing.

XIAP: X-Linked Inhibitor of Apoptosis

This gene encodes for a protein that protects immune cells from self- destructing by blocking the action of caspases.

XK: X-Linked Kx Blood Group

This gene encodes for a protein that is present on the surface of red blood cells and carries a molecule known as the Kx blood group antigen.

trans-resveratrol-3-O-sulfate reduced C/EBP alpha expression. Thus, the suppression of C/EBP by these natural products resulted in reduced regulation of genes involved in immune and inflammatory responses.

Yet another study revealed that wolfberry phytochemicals, including rutin, activate C/EBP-homologous protein and may increase immune and inflammatory responses. Apigenin reduces the TG- and BFA-induced expression of the EBP homologous protein. CAPE suppresses 3T3-L1 differentiation to adipocytes via the inhibition of C/EBP alpha. Hesperidin differentially influenced C/EBP binding on the CYP19 promoter. Quercetin inhibited ER stress C/EBP homologous protein pathway-mediated apoptosis. Ferulic acid protective against ER stress C/EBP homologous protein induced neuronal cell death and may provide a possible new treatment for Alzheimer's disease. Carnosic acid upregulates the ER stress C/EBP homologous protein and induced the cell death in resistant myeloid leukemia cells.

Muscular dystrophies (DMD) is caused by mutations of the DMD gene. Studies have shown that several natural products can be beneficial in the treatment of DMD. Curcumin inhibits the NF-kappaB activity, which can be useful in the therapy of DMD in mdx mice. Ascorbic acid significantly upregulated the α6 chain in DMD patients. In addition, genistein treatment has several beneficial effects on DMD patients. The genistein treatment caused increased forelimb strength, decreased serum creatine-kinase levels, reduced markers of oxidative stress, lessened muscle necrosis, and enhanced regeneration. Resveratrol could serve as a therapy for DMD by reducing inflammation. Moreover, green tea extract could be advantageous in the treatment of DMD patients. Clearly, numerous natural products have a significant influence on genes in chromosome X, and the interactions of these selected natural products with other genes of chromosome X have not been discussed in this chapter.

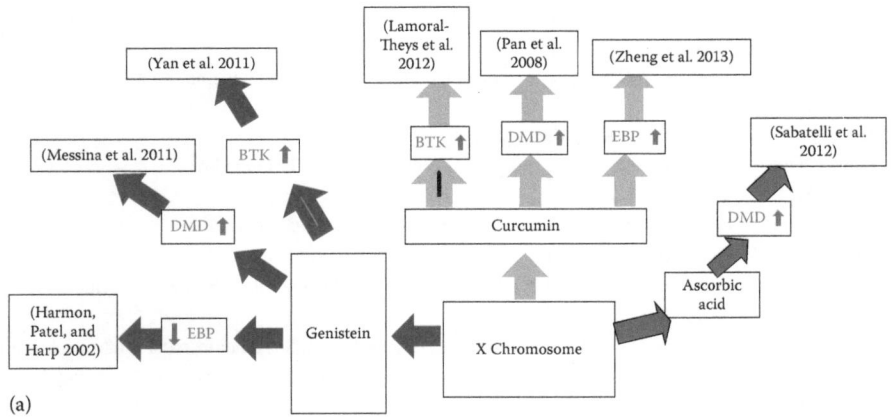

(a)

FIGURE 2.23   (See color insert.) Natural products interacting with the selected genes of chromosome X. (a) BTK [BTK and curcumin]: http://www.ncbi.nlm.nih.gov/pubmed/?term=BTK+AND+Curcumin; BTK [BTK and genistein]: http://www.ncbi.nlm.nih.gov/pubmed/?term=BTK++AND+Genistein; DMD [DMD and ascorbic acid]: http://www.ncbi.nlm.nih.gov/pubmed/?term=DMD+AND+Acid+ascorbic; DMD [DMD and curcumin]: http://www.ncbi.nlm.nih.gov/pubmed/?term=DMD+AND+Curcumin; DMD [DMD and genistein]: http://www.ncbi.nlm.nih.gov/pubmed/?term=DMD+AND+Genistein; EBP [EBP and curcumin]: http://www.ncbi.nlm.nih.gov/pubmed/?term=EBP+AND+Curcumin; and EBP [EBP and genistein]: http://www.ncbi.nlm.nih.gov/pubmed/?term=EBP+AND+Genistein. (Data from Harmon, A. W. et al., *Biochem J*, 367, 203–8, 2002; Lamoral-Theys, D. et al., *J Cell Mol Med*, 16, 1421–34, 2012; Messina, S. et al., *Neuromuscul Disord*, 21, 579–89, 2011; Pan, Y. et al., *Mol Cells*, 25, 531–7, 2008; Sabatelli, P. et al., *Matrix Biol*, 31, 187–96, 2012; Yan, G. R. et al., *J Proteomics*, 75, 695–707, 2011; Zheng, M. et al., *Int Immunopharmacol*, 15, 517–23, 2013.)   (*Continued*)

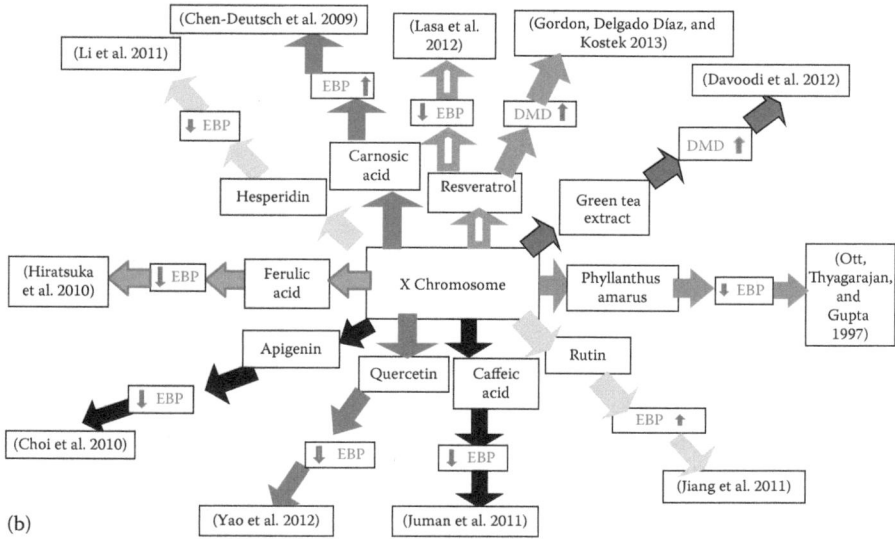

FIGURE 2.23 (Continued)   Natural products interacting with the selected genes of chromosome X. (b) DMD [DMD and green tea extract]: http://www.ncbi.nlm.nih.gov/pubmed/?term=DMD+AND+Green+tea+extract; DMD [DMD and resveratrol]: http://www.ncbi.nlm.nih.gov/pubmed/?term=DMD+AND+Resveratrol; EBP [EBP and apigenin]: http://www.ncbi.nlm.nih.gov/pubmed/?term=EBP+AND+Apigenin; EBP [EBP and caffeic acid]: http://www.ncbi.nlm.nih.gov/pubmed/?term=EBP+AND+Caffeic+acid; EBP [EBP and carnosic acidl]: http://www.ncbi.nlm.nih.gov/pubmed/?term=EBP+AND+Carnosic+acidl; EBP [EBP and ferulic acid]: http://www.ncbi.nlm.nih.gov/pubmed/?term=EBP+AND+Ferulic+acid; EBP [EBP and hesperidin]: http://www.ncbi.nlm.nih.gov/pubmed/?term=EBP+AND+Hesperidin.; EBP [EBP and phyllanthus amarus]: http://www.ncbi.nlm.nih.gov/pubmed/?term=EBP+AND+Phyllanthus+amarus; EBP [EBP and quercetin]: http://www.ncbi.nlm.nih.gov/pubmed/?term=EBP+AND+Quercetin; EBP [EBP and resveratrol]: http://www.ncbi.nlm.nih.gov/pubmed/?term=EBP+AND+Resveratrol; and EBP [EBP and rutin]: http://www.ncbi.nlm.nih.gov/pubmed/?term=EBP+AND+Rutin. (Data from Chen-Deutsch, X. et al., *Leuk Res*, 33, 1372–8, 2009; Choi, A. Y. et al., *Neurochem Int*, 57, 143–52, 2010; Davoodi, J. et al., *Phys Med Rehabil Clin N Am*, 23, 187–99, xii–xiii, 2012; Gordon, B. S. et al., *Clin Nutr*, 32, 104–11, 2013; Hiratsuka, T. et al., *PLoS One*, 5, e13280, 2010; Jiang, Y. et al., *J Nutr Food Sci*, S2, 2011; Juman, S. et al., *Biol Pharm Bull*, 34, 490–4, 2011; Lasa, A. et al., *Mol Nutr Food Res*, 56, 1559–68, 2012; Li, F. et al., *Mol Cell Endocrinol*, 344, 51–8, 2011; Ott, M. et al., *Eur J Clin Invest*, 27, 908–15, 1997; Yao, S. et al., *Exp Biol Med (Maywood)*, 237, 822–31, 2012.)

2.25 CHROMOSOME Y

Table 2.47 summarizes six types of genetic diseases from chromosome Y. Table 2.48 provides the list of genes that are chosen for the natural products interactions studies. There are no reports in PubMed about the selected genes and natural products interactions. Research on the effects of natural products on the genes of the Y chromosome may still be in the process of taking place or have yet to be performed.

TABLE 2.47
Genetic Diseases Linked to Chromosome Y

1. Y chromosome microdeletion (YCM)
2. Defective Y chromosome
3. Klinefelter's syndrome (47, XXY)
4. 47,XYY syndrome
5. More than two Y chromosomes
6. XX male syndrome

TABLE 2.48
Chromosome Y Selected Genes

AZF1: Azoospermia Factor 1

This gene is involved in spermatogenesis in the testes.

BPY2: Basic Protein on the Y Chromosome

This gene encodes for a protein that interacts with ubiquitin protein ligase E3A and might be involved in the male germ cell development and infertility.

DAZ1: Deleted in Azoospermia 1

This gene's expression is restricted to pre-meiotic germ cells, mostly in spermatogonia, and it encodes for an RNA-binding protein that is important for spermatogenesis.

PRKY: Protein Kinase, Y-Linked, Pseudogene

This gene has lost a coding exon that results in all transcripts being candidates for nonsense-mediated decay and unlikely to express a protein.

RBMY1A1: RNA-Binding Motif Protein, Y-Linked, Family 1, Member A1

This gene encodes for a protein that contains an RNA-binding motif in the N-terminus and four SRGY boxes in the C-terminus.

SRY: Sex-Determining Region

This gene has no introns, and it encodes for a transcription factor that is part of the high motility group (HMG)-box family of DNA-binding proteins.

TSPY: Testis-Specific Protein, Y-Linked 1

This gene encodes for a protein that is found only in testicular tissue and may be involved in spermatogenesis.

USP9Y: Ubiquitin-Specific Peptidase 9, Y-Linked

This gene encodes for a protein that is similar to ubiquitin-specific proteases, which cleave the ubiquitin moiety from ubiquitin-fused precursors and ubiquitinylated proteins.

UTY: Ubiquitously Transcribed Tetratricopeptide Repeat Containing, Y-Linked

This gene encodes for a protein that contains tetratricopeptide repeats, which are thought to be involved in protein–protein interactions.

ZFY: Zinc-Finger Protein, Y-Linked

This gene encodes for a protein that may act as a transcription factor.

2.26   CONCLUSION

When we study these natural products interactions with genes of different chromosomes, we find that there are several genes interacting with one compound and many natural products interacting with one gene either up or down to regulate the gene activity (Table 2.49). Also, it is interesting to note why the particular natural product is interacting with genes of chromosomes of specific tissues. If we identify the unique nature of the structure of the natural product, then it is easy to compare similar structural compounds on this gene activity in vitro. We urge the readers to refer to the specific gene interactions by looking into the original article. We randomly scanned these 28 natural products on selected genes of human chromosomes and tabulated the status of the interactions (Table 2.50).

Among the selected genes for interactions, we presented only the interacting genes, and other genes are not available in the literature or the research is yet to be done on these compounds. From Table 2.49, it is inferred that most of the natural products exhibit an increase or decrease in specific gene activity directly or indirectly among 72 genes studied by scanning the literature. However, there are few natural products that have different roles on a particular gene. They are as follows:

1. IP-10 gene activity is decreased by quercetin but increased by green tea.
2. ERG gene activity is enhanced by genistein but decreased by resveratrol.
3. VEGFA gene function is decreased by genistein, resveratrol, ascorbic acid, apigenin, caffeic acid, curcumin, quercetin, and green tea but increased by ferulic acid.
4. StAR gene activity is decreased by genistein but increased by apigenin and quercetin.
5. TH gene activity is decreased by apigenin and ferulic acid but increased by genistein and naringin.
6. RELA gene activity is increased by resveratrol but decreased by genistein, rosmarinic acid, caffeic acid, curcumin, and rutin.
7. BDNF gene activity is increased by genistein, caffeic acid, and curcumin but decreased by quercetin and ferulic acid.
8. CASP3 gene activity is increased by resveratrol, caffeic acid, curcumin, and hesperidin but decreased by carnosic acid, quercetin, ferulic acid, rutin, and lycopene.
9. RB1 gene activity is decreased by genistein, apigenin, lycopene, ferulic acid, and hesperidin but increased by ascorbic acid.
10. GFAP gene activity is decreased by curcumin, quercetin, ferulic acid, hesperidin, and rutin but increased by resveratrol and chrysin.
11. APOE gene activity is increased by caffeic acid but decreased by hesperidin.
12. COMT gene activity is increased by genistein, resveratrol, ascorbic acid, and green tea but decreased by quercetin and diosmin.
13. EP300 gene activity is increased by genistein but decreased by resveratrol and curcumin.
14. EBP gene activity is decreased by apigenin, resveratrol, caffeic acid, hesperidin, and quercetin but increased by genistein, curcumin, and rutin.

TABLE 2.49
Natural Products Effects on Specific Gene in Human Chromosomes

Chrom/ Genes #	Gene Name	Genistein	Resveratrol	Ascorbic Acid	Rosmarinic Acid	Apigenin	Caffeic Acid	Carnosic Acid	Curcumin	Quercetin	Green Tea (EGCG)	Lycopene	Ferulic Acid	Hesperidin	Rutin	Chrysin	Diosmin	Naringin	Thymol
Chrom 1																			
1	NKCC1	Increases								Increases									
2	NGF	Increases	Increases	Increases	Increases	Increases	Increases	Increases	Increases	Increases									
3	trkA and EGR-1		Increases																
4	COL11 A1			Increases															
Chrom 2																			
1	MSH2								Increases	Increases									
2	TPO	Decreases				Decreases													
3	COL3A1			Increases															
Chrom 3																			
1	PLD1										Increases								
2	GATA2								Decreases										
3	PIK3CA										Increases								
Chrom 4																			
1	IP-10									Decreases	Increases								
2	MIP-2						Decreases		Decreases	Decreases		Decreases							
3	sFRP2	Increases																	
Chrom 5																			
1	APC	Increases			Increases		Increases	Increases	Increases	Increases	Increases								
2	EGR	Increases	Decreases						Increases	Increases									

(Continued)

TABLE 2.49 (Continued)
Natural Products Effects on Specific Gene in Human Chromosomes

Chrom/ Genes #	Gene Name	Genistein	Resveratrol	Ascorbic Acid	Rosmarinic Acid	Apigenin	Caffeic Acid	Carnosic Acid	Curcumin	Quercetin	Green Tea (EGCG)	Lycopene	Ferulic Acid	Hesperidin	Rutin	Chrysin	Diosmin	Naringin	Thymol
Chrom 6																			
1	RUNX2	Increases	Increases	Increases			Increases		Increases	Increases				Increases					
2	VEGFA	Decreases	Decreases	Decreases		Decreases	Decreases		Decreases	Decreases	Decreases		Increases						
Chrom 7																			
1	MyoD													Increases					
2	IL6 gene										Decreases								
3	P-gp	Increases													Increases				
4	CYP3A4														Increases				
5	eNOS	Increases	Increases						Increases	Increases				Increases					
Chrom 8																			
1	StAR	Decreases				Increases				Increases									
2	LPL									Increases									
3	IDO				Decreases														
Chrom 9																			
1	TLR4		Decreases				Decreases		Decreases				Decreases						
Chrom 10																			
1	SIRT1	Increases	Increases						Increases	Increases									
Chrom 11																			
1	TH	Increases				Decreases			Decreases			Decreases	Decreases					Increases	
2	CPT1A	Increases																	
3	RELA	Decreases	Increases		Decreases		Decreases		Decreases						Decreases				
4	BDNF	Increases	Increases				Increases		Decreases	Increases			Increases						

(Continued)

TABLE 2.49 (Continued)
Natural Products Effects on Specific Gene in Human Chromosomes

Chrom/ Genes #	Gene Name	Genistein	Resveratrol	Ascorbic Acid	Rosmarinic Acid	Apigenin	Caffeic Acid	Carnosic Acid	Curcumin	Quercetin	Green Tea (EGCG)	Lycopene	Ferulic Acid	Hesperidin	Rutin	Chrysin	Diosmin	Naringin	Thymol
5	WT-1								Decreases										
6	PKC-α								Decreases										
7	IL-18		Decreases						Decreases										
8	CASP3	Increases					Increases	Decreases	Increases	Decreases		Decreases	Decreases	Increases	Decreases	Decreases	Decreases		
9	ACAT1													Decreases					
Chrom 12																			
1	CDK2								Decreases	Decreases		Decreases		Decreases		Decreases		Decreases	
2	UBC								Increases	Decreases									
3	KERA			Increases															
4	LRRK2			Increases					Increases										
5	COL2A1			Increases															
6	PAH			Increases															
7	KRAS																		
Chrom 13																			
1	Rb1	Decreases		Decreases		Decreases						Decreases			Increases			Increases	
2	BRCA2							Increases			Increases	Increases	Decreases	Decreases				Increases	Increases
Chrom 14																			
1	GCH1	Increases																	
2	Ataxin-3													Increases					

(Continued)

TABLE 2.49 (Continued)
Natural Products Effects on Specific Gene in Human Chromosomes

Chrom/ Genes #	Gene Name	Natural Products																	
		Genistein	Resveratrol	Ascorbic Acid	Rosmarinic Acid	Apigenin	Caffeic Acid	Carnosic Acid	Curcumin	Quercetin	Green Tea (EGCG)	Lycopene	Ferulic Acid	Hesperidin	Rutin	Chrysin	Diosmin	Naringin	Thymol
Chrom 15																			
1	Furin									Decreases									Decreases
2	IGF1R					Decreases				Decreases					Decreases				
3	PML	Increases																	
4	B2M								Increases										
Chrom 16																			
1	MMP-2							Decreases		Decreases	Decreases								
2	MC1R			Decreases							Decreases								
Chrom 17																			
1	NF1						Decreases												
2	GFAP	Increases					Increases		Decreases	Decreases			Decreases	Decreases	Increases				
3	BRCA1	Increases	Increases						Increases			Increases							
1	GRP		Increases			Decreases													
2	MBP									Decreases									
3	MAPK4						Decreases												
Chrom 19																			
1	PRX			Decreases				Decreases		Decreases									
2	APOE						Increases							Decreases					
3	STK11		Increases																
Chrom 20																			
1	PRNP	Increases	Increases																
2	BMP-2		Increases						Increases					Increases				Increases	

(*Continued*)

TABLE 2.49 (Continued)
Natural Products Effects on Specific Gene in Human Chromosomes

Chrom/ Genes #	Gene Name	Genistein	Resveratrol	Ascorbic Acid	Rosmarinic Acid	Apigenin	Caffeic Acid	Carnosic Acid	Curcumin	Quercetin	Green Tea (EGCG)	Lycopene	Ferulic Acid	Hesperidin	Rutin	Chrysin	Diosmin	Naringin	Thymol
Chrom 21																			
1	KCNE1	Decreases																	
2	SOD1		Increases			Increases	Increases	Increases											
3	APP					Decreases	Decreases	Decreases	Decreases					Decreases		Decreases			
Chrom 22																			
1	COMT	Increases	Increases	Increases						Decreases	Increases							Decreases	
2	EP300		Decreases						Decreases										
3	SOX10			Increases															
Chrom X																			
1	BTK	Increases																	
2	EDA		Decreases																
3	EBP	Increases	Decreases			Decreases	Decreases		Increases	Decreases				Decreases	Increases				
4	DMD	Increases		Increases					Increases	Increases	Increases								

TABLE 2.50

Total Number of Selected Genes Interacting with Natural Products

Name of the Natural Products	Number of Selected Genes Interacting
1. Curcumin	27
2. Genistein	25
3. Quercetin	23
4. Resveratrol	22
5. Caffeic acid	16
6. Hesperidin	13
7. Ascorbic acid	12
8. Apigenin	11
9. Green tea (EGCG)	9
10. Rutin	8
11. Ferulic acid	7
12. Carnosic acid	6
13. Lycopene	5
14. Chrysin	4
15. Rosmarinic acid	4
16. Naringin	4
17. Diosmin	2

The different roles of the natural products on certain genes indicate that caution should be exercised when considering a combinatorial therapeutic approach for the specific gene-based disease. In Chapter 3, we present the gene activity profile for one natural product, citrus limonin, on the genes involved in metastasis of breast cancer in an animal model.

ACKNOWLEDGMENTS

The data analysis help was provided by Mary Adkins, Whitney Backhaus, Swathi Ariyapadi, Kelsey Junek, Jason Chandler, and Nardos Gossa.

3 Interaction of Citrus Natural Products with Genomes

Siva G. Somasundaram, Janet Price,
G. K. Jayaprakasha, and Bhimanagouda S. Patil

CONTENTS

3.1 INTERACTION OF CITRUS NATURAL PRODUCTS WITH ANTI-INFLAMMATORY AND ANTIATHEROGENIC GENOME

3.1.1 INTRODUCTION

The exposure of populations worldwide to citrus fruits, and its many uses, has led to studies aimed at elucidating some of its interaction with genome activities. Citrus limonin and related compounds inhibit free radical generation and act as free radical scavengers. This property is beneficial in preventing the free radical damage to the genome, especially the histones and DNA itself from mutations. In addition, they also influence the blood cells and immune cells to protect from the pathogenic organisms by altering the activities of genes.

3.1.2   NUTRIGENOMIC STUDIES OF CITRUS JUICE AND ITS NATURAL PRODUCT HESPERIDIN

In general, 444 mg/L hesperidin is present in commercial orange juices (Manach et al., 2003). The nutrigenomic profile of orange juice and its ingredient hesperidin was studied extensively by Milenkovic et al. (2011). The clinical trial (Trial Registration: ClinicalTrials.gov NCT 00983086) study shows that 28 days regular consumption of orange juice or hesperidin alone changes the blood leukocyte gene expression from normal to an anti-inflammatory and antiatherogenic genomic profile.

The findings indicate that the expression of 3422 genes has been changed by orange juice consumption. Citrus natural product hesperidin alters the expression of 1819 genes. The genomic expression profile changes for 1582 genes are common for both orange juice and hesperidin group (Figure 3.1). Figure 3.2 demonstrates that the major genomic interactions are centered on the interactions of nuclear factor kappa B (NF-κB) gene transcription factor. The change of genomic profiles includes the physiological functions of chemotaxis, adhesion, infiltration, and lipid transport and helps in lowering the lipid accumulation in the vascular wall.

The CX3R1 gene is expressed more and involved in recruitments of monocytes in the atherosclerotic arteries and not in the normal artery (Wong, Wong, and McManus, 2002). These chemokines are chemoattractants for the activation of leukocytes (Weber, Schober, and Zernecke, 2004). This genomic profile study (Milenkovic et al., 2011) proved that the genes that are downregulated by the citrus natural products specifically involved in the atherosclerosis process are coding for chemokines such as CCL26 and

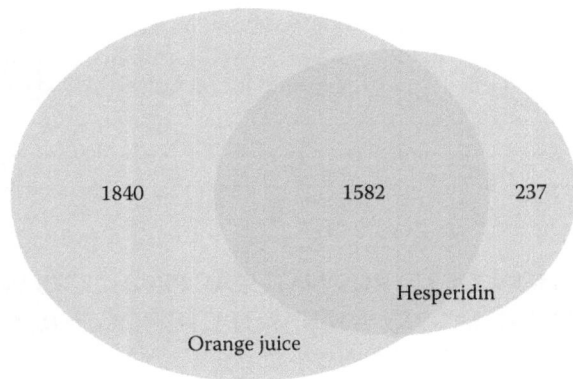

FIGURE 3.1   (See color insert.) A Venn diagram of the number of differentially expressed genes in human blood leukocytes after a 4-week consumption of orange juice, hesperidin, or both: Expression of these genes was significantly different from the control drink consumption group. (Data from Milenkovic, D. et al., *PLoS One*, 6, e26669, 2011.)

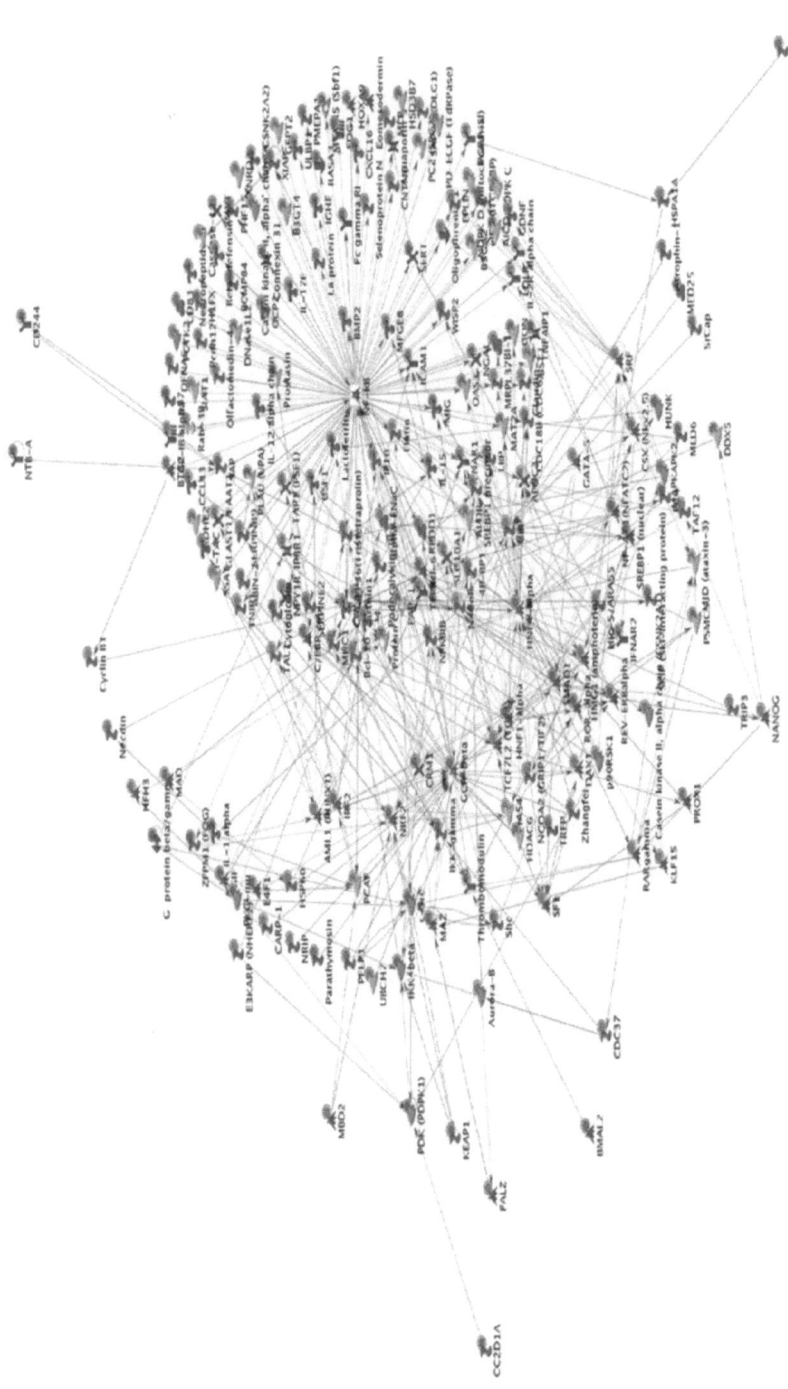

FIGURE 3.2 (See color insert.) A representative biological network based on the differentially expressed genes of the orange juice group using MetaCore™ network software and the analyze network algorithm: The network shown is the NF-κB network. Dots in the right corner of a gene indicate differential expression. (Data from Milenkovic, D. et al., *PLoS One*, 6, e26669, 2011.)

CX3R1. Apart from these genes, citrus natural products further downregulate other genes such as CCL26, IL14, MCP-1, CCL2, CXCL11, and CXCL16 (Apostolakis et al., 2006; Damås et al., 2007). During atherosclerosis, these genes are highly activated through the NF-κB gene (Hishikawa, Nakaki, and Fujita, 2005). It is interesting to note that citrus compounds upregulate the inhibitor of NF-κB, such as NF-κBIB, and thereby inhibit the upregulation of these genes involved in atherosclerosis (Figure 3.2). In addition, these compounds downregulate the gene CD80 through the downregulation of transcriptional factor BCL6, which is highly expressed in cardiovascular arteries (Wang et al., 2008).

Further, citrus juices and hesperidin increase the expression of a cell adhesion molecule CEACAM3 that is negatively involved in neutrophil adhesion to the endothelial cells (Skubitz and Skubitz, 2008). An integrin gene such as ITGBL1 is decreased by both orange juice and hesperidin. However, orange juice decreased other integrin genes such as ITGA5, ITGA7, and ITGAX. It is interesting to note in the following study that in the breast cancer metastasis genome, citrus limonin activates the integrin beta 3 gene. Connexin 31.3 gene is downregulated by orange juice and hesperidin. Orange juice alone downregulated more connexin genes such as connexins 30, 40, and 46. Also, the LDL receptor and ACAT2 genes are downregulated by the citrus natural products and help in the prevention of cardiovascular disease. Another study from hesperidin supports that this flavanone-rich diet reduces the risk of coronary heart disease (Mink et al., 2007).

Similarly, a genomic study comes from the interactions of a different natural product, quercetin, with the human monocyte gene expression. The dosage used was 150 mg quercetin or placebo daily for six weeks, and the authors found that four genes were significantly changed when compared to the placebo-treated group. These four genes include C1GALT1, involved in O-glycan biosynthesis; GM2A, involved in glycolipid catabolism; HDGF, involved in cell proliferation; and SERPINB9, involved in apoptosis (Boomgaarden et al., 2010). Then, we would like to pursue the following study to check whether the citrus natural product limonin has any metastasis genomic interactions due to its impact on genes of vascular tissues.

3.2 INTERACTION OF CITRUS NATURAL PRODUCTS WITH CANCER METASTASIS GENOME

In the following study, we focus on the genomic interaction profile of citrus limonin with breast cancer metastasis. Citrus natural products such as limonin and hesperidin interfere with reactive oxygen species metabolism and can act as potential candidates for the treatment of cancer. Some of the glucosides of citrus limonin inhibit the endogenously generated reactive oxygen species. Our previous studies (Somasundaram et al., 2012) demonstrated the interactions of limonin with chemotherapy in breast cancer cell lines MCF-7 (p53 wild type) and MDA-MB-231 (p53 mutant) as well as the nontumorigenic epithelial cell line MCF-10 to study the gene activities of NF-κB, p38 and ERK-MAPK signaling kinases. In addition, the interacting effect of cyclophosphamide and limonin on MDA-MB-231 xenografts was also studied. Both in vitro and in vivo results suggest that the citrus natural product limonin could be beneficial for breast cancer patients undergoing chemotherapy.

The next study from our group indicates the interacting effect of signaling pathway involved in genes of tumor progression and metastasis. Citrus-derived limonoids are reported to act against cancer cells in various ways (Tian et al., 2001; Patil et al., 2009). In a rat model of azoxymethane-induced colon cancer (aberrant crypt foci formation), limonoids were shown to suppress proliferation and enhance apoptosis by lowering levels of cyclooxygenase-2 (COX-2) and inducible nitric oxide synthase (iNOS) induced by the carcinogen (Vanamala et al., 2006), potentially via the NF-κB pathway. This signaling pathway can regulate many genes that are important for tumor progression and metastasis (Singh and Aggarwal, 1995). Further, our study reports that, unlike curcumin, limonin lacks antichemotherapeutic potentials in breast cancer xenografts.

However, the role of limonin on metastasis-associated gene expression has not been studied. From the earlier genomic signature reviewed by Urquidi and Goodison (2007), it has been demonstrated that the systemic adjuvant therapy given to lymph node (LN) negative patients provides an opportunity to study the risk of recurrence. We explore the possibility of involvement of dietary limonin as an adjuvant on differentially expressed genes associated with LN metastasis in an experimental breast cancer cell line. The present study used two isogenic human breast cancer cell lines, which differ in ability to grow and metastasize in immunodeficient mice (Lev, Kiriakova, and Price, 2003). The cell lines represent the triple-negative class of breast cancers, lacking expression of the hormone receptors such as estrogen receptor (ER) and progesterone receptor (PR), and with low expression of human epidermal growth factor receptor 2 (HER2).

Differential expression of a panel of genes by the two cell lines has been previously reported. High expression of several of the differentially expressed genes was significantly associated with LN metastasis and poor prognosis when analyzed in tissue microarrays of primary breast cancers (Chelouche-Lev et al., 2004; Kluger et al., 2005). Gene expression patterns observed by microarrays help to distinguish the molecular subtypes of breast cancer that are used to predict metastatic relapse (Sørlie et al., 2001; van de Vijver et al., 2002; Weigelt et al., 2003, 2005). Thus, studying the metastasis-associated genes expressed by the established cell lines may have some relevance for the malignant progression of human breast cancer. This pilot study was initiated to test whether exposure to citrus-derived limonin modulated expression of metastasis-associated genes in the breast cancer cell lines. The results reveal the importance of the nutrigeomics approach to the metastastic genome in breast cancer cell lines.

3.2.1 Methods and Materials

3.2.1.1 Cell Lines

The GI101A human breast cancer cell line and the GILM2 variant of this line, derived as reported previously (Kluger et al., 2005), were maintained in a monolayer culture in Dulbecco's minimum essential medium (4 mM glucose), supplemented with 10% fetal calf serum, L-glutamine, and penicillin–streptomycin. The BT-20 and MDA-MB-231 human breast cancer cell lines were maintained in Eagle's minimum essential medium, supplemented with 5% fetal calf serum, L-glutamine, nonessential amino acids, sodium pyruvate, MEM vitamins, penicillin (100 U/mL) and streptomycin (100 µg/mL). All cell lines were tested for and found free of *Mycoplasma* sp. infection.

3.2.1.2   In Vitro Responses to Limonin

Limonin was isolated and purified as per the established method (Mandadi et al., 2007). Tumor cells were plated in 96-well culture plates at an initial density of 1×10^4 cells per well and allowed to attach for 24 h. The culture medium was then changed to medium with different concentrations of limonin from a stock solution prepared in dimethyl sulfoxide (DMSO) at a concentration of 100 mM. Some cultures were treated with curcumin, which was also dissolved in DMSO. The control condition was medium with an equivalent concentration of DMSO as the highest concentration of limonin or curcumin. The cells were incubated for up to 72 h; the relative cell numbers were assessed using MTT. The conversion of MTT to formazan in metabolically viable cells was monitored with a MR-5000 microtiter plate reader set to read at 570 nm.

For two cell lines, MDA-MB-231 and GILM2, protein lysates were collected from cultures exposed to limonin for 6 h to determine whether the agent altered activation of the MAPK/ERK1/2 signaling pathway. Cells were plated in six-well plates, cultured in serum-depleted medium (0.5% serum) for 24 h, and then exposed to limonin (100 µM), DMSO (0.1%), or the MEK inhibitor U0126 (10 µM) in serum-free medium. After 6 h of treatment with these agents, the cultures were stimulated with the addition of EGF (50 ng/mL), and total protein lysates were prepared after 10 min. Previous studies of 24 h exposure to limonin at lower concentrations, 1, 5, and 10 µM, showed it did not have any profound effect on the increase of phosphorylations of ERK in MDA-MB-231 cells. Hence, in the present study, we use higher concentrations with a reduced time (6 h) of exposure to check the phosphorylation of ERK in MDA-MB-231 cells. Aliquots of 20 µg protein were separated on polyacrylamide gels and transferred to nitrocellulose membranes. Antibodies used for immunoblotting work against phospho-p44/p42 MAP kinase (Thr202/Tyr204) and ERK1/2 (Cell Signaling Technology, Inc., Beverly, MA) were detected using a horseradish peroxidase-conjugated anti-rabbit IgG and Amersham's ECL system (Amersham, Arlington Heights, IL).

3.2.1.3   In Vitro Migration Assays

Assays in Transwell chambers were used to evaluate whether limonin could modulate the migration potential of cancer cells. Suspensions of GI101A and GILM2 cells in serum-free medium were plated into the upper well of the chambers over membranes with 8 µM pores. The cells were mixed with 0.025% DMSO, 25 µM limonin, or 1 µM of an Akt-inhibitor (Akt-inhibitor-IV, EMD Bioscience, Billerica, MA). The chemoattractant in the lower chamber was medium supplemented with 1% fetal calf serum. After 18 h incubation, the cell suspensions were aspirated from the upper chambers, and cells attached to the upper surface of the membrane were removed by wiping with cotton swabs. The membranes were fixed in methanol and stained with hematoxylin and eosin. Cells that had migrated to the underside of the membrane were counted in representative fields, at least 10 fields per filter in replicate samples.

3.2.1.4 Gene Expression Assays

The GI101A and GILM2 cells were plated into 100 mm plates in Dulbecco's modified Eagle medium (DMEM) with 10% fetal calf serum. The following day when the cultures were of approximately 80% confluence, the culture medium was replaced with medium supplemented with either 0.025% DMSO or 25 μM limonin. The cultures were incubated for a further 18 h, and then the cells were collected for isolation of the total RNA using the ArrayGrade Total RNA Isolation Kit (SABiosciences Corp., Frederick, MD). The RNA samples were applied to RT^2 Profiler PCR Arrays from SABiosciences Corp. with the 84 genes known to be involved in metastasis. RNA was collected from two sets of treated cultures, and each applied to four replicate plates. The real-time polymerase chain reaction (PCR) was performed using an ABI 7500 instrument (Applied Biosystems, Foster City, CA), and the data were analyzed using a web portal provided by SABiosciences. Results of the analyses were downloaded as Excel files.

Real-time PCR was performed for selected genes using the same total RNA samples, reverse-transcribed with random primers from the High Capacity cDNA Archive Kit (Applied Biosystems). cDNA was amplified in duplicate samples using Predeveloped TaqMan Assay Reagents for VEGF, IL-8, CXCL1, CXCL12, and 18S, following the manufacturer's recommended amplification procedure. Results were recorded as mean threshold cycle (Ct), and relative expression was determined using the comparative Ct method. The ΔCt was calculated as the difference between the average Ct value of the endogenous control (18S) and the average Ct value of the test gene. To compare the relative amount of target gene expression in different samples, human placenta RNA was used as a calibrator. The ΔΔCt was determined by subtracting the ΔCt of the calibrator from the ΔCt of the test sample. Relative expression of the target gene was calculated by the formula $2^{-\Delta\Delta Ct}$, which is the amount of gene product normalized to the endogenous control relative to the calibrator sample.

3.2.1.5 Statistical Analyses

Analysis of variance, with Bonferroni's multiple comparison test and Student's t-test, was performed to determine the statistical significance.

3.2.2 RESULTS

3.2.2.1 In Vitro Sensitivity of Breast Cancer Cells to Limonin

The MTT assay was used to evaluate the response of breast cancer cells to limonin over a range of doses from 0.5 to 150 μM. Higher concentrations of limonin were not used as the agent came out of solution in the culture medium at solutions above 150 μM. The data in Figures 3.1 through 3.3 show that the different breast cancer cell lines show minimal response to limonin. The greatest statistically significant response was seen in the MDA-MB-231 cell line with a 25% reduction ($p < 0.05$) at the 150 μM dose (Figure 3.3). In contrast, the cell lines were sensitive to curcumin with a marked reduction in cell numbers at concentrations greater than 10 μM (shown for GI101A and GILM2 in Figure 3.4, with data from other cell lines not shown).

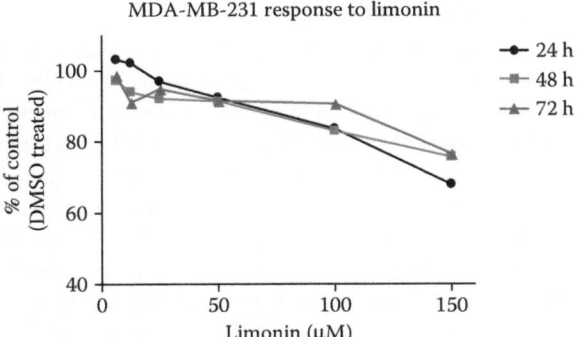

FIGURE 3.3 The results of MTT assays for relative cell numbers in cultures of MDA-MB-231 cells treated with different concentrations of limonin, and expressed as a percentage of the control, DMSO-treated cells: Replicate plates of cells were analyzed at 24, 48, and 72 h after the addition of DMSO or limonin.

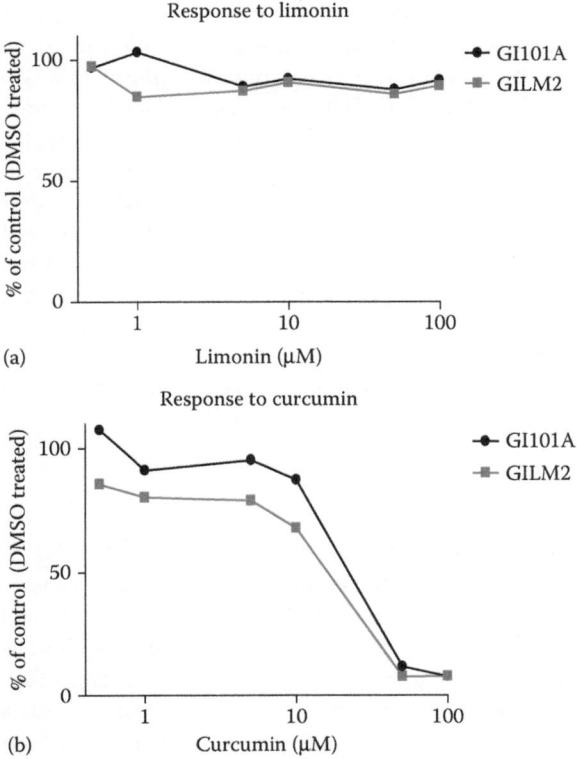

FIGURE 3.4 The results of MTT assays for relative cell numbers in cultures of GI101A and GILM2 cells treated with different concentrations of either limonin (a) or curcumin (b), and expressed as a percentage of the control DMSO-treated cells: The MTT assays were performed 48 h after the addition of DMSO, limonin, or curcumin.

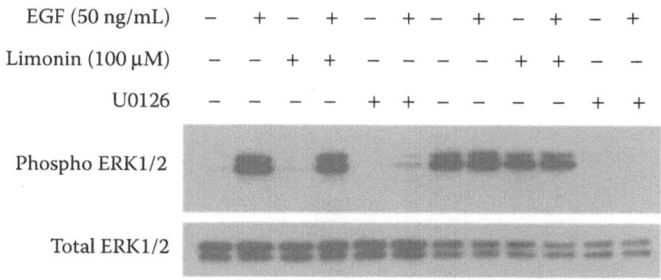

| EGF (50 ng/mL) | – | + | – | + | – | + | – | + | – | + | – | + |

Phospho ERK1/2

Total ERK1/2

FIGURE 3.5 Phosphorylation of ERK1/2 in breast cancer cells. GILM2 and MDA-MB-231 cells were incubated with 100 μM limonin or 10 μM U0126 for 6 h before stimulation with 50 ng/mL EGF for 10 min before cell lysis to collect proteins: Phosphorylation of ERK1/2 was detected by immunoblotting using phospho-specific antibody. The filter was then re-probed with an antibody that recognizes all forms of the ERK1/2 protein (Kd 42/44).

Figure 3.5 shows the results of immunoblotting of protein lysates collected from MDA-MB-231 and GILM2 cells, for the activation of ERK1/2. The MDA-MB-231 cell line carries a mutant K-Ras, with constitutively activated ERK1/2, and signal was seen in both serum-starved and EGF-treated samples. The samples exposed to the MEK inhibitor U0126 exhibited an inhibition of ERK1/2 phosphorylation. The activation of ERK1/2 was stimulated in GILM2 cells treated with EGF and was blocked by U0126. The addition of limonin to the cultures neither enhanced nor attenuated the activation of ERK1/2 in either of the breast cancer cell lines.

3.2.2.2 In Vitro Migration Assays

The motility of the GI101A and GILM2 cells was assessed by scoring the number of cells that had migrated through the 8.0 μM pores of membranes in a Transwell chamber, with 1.0% fetal calf serum as stimulant. As shown in Figure 3.6a, a representative result from three independent experiments, the GILM2 variant line has greater motility than the GI101A cell line. The migration of both cell lines was significantly inhibited in the presence of the Akt inhibitor; the Akt signaling pathway has been shown by others to contribute to the migration of cells (Onishi et al., 2007). The addition of 25 μM limonin had no effect on the ability of the cells to migrate. Figure 3.6b represents the average fold change in migration from three experiments. Comparing the fold-differences between DMSO-treated cells and the Akt inhibitor, the migration of GI101A cells was reduced by 42% ($p = 0.012$) and the migration of GILM2 cells was reduced by 39% ($p = 0.011$, student's t-test).

3.2.3 Discussion

Citrus-derived limonin is reported to have various activities that may affect the survival and proliferation of cancer cells, although the molecular mechanisms involved have not been fully identified (Tian et al., 2001; Vanamala et al., 2006; Patil et al., 2009). This study sought to demonstrate an effect of limonin on human breast cancer cells, using assays for growth and migration and comparisons of

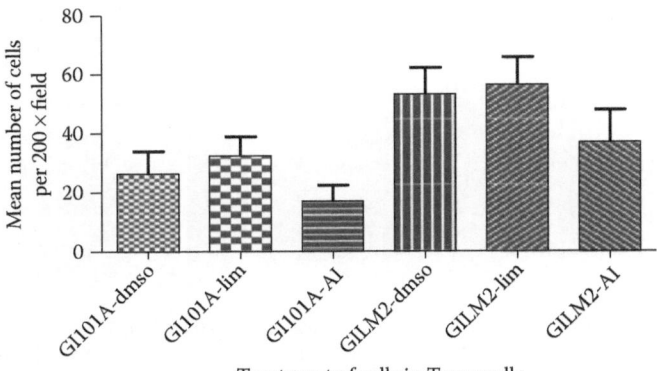

DMSO = 0.025%
Limonin = 25 μM
AI (Akt inhibitor IV) = 1 μM
(a)

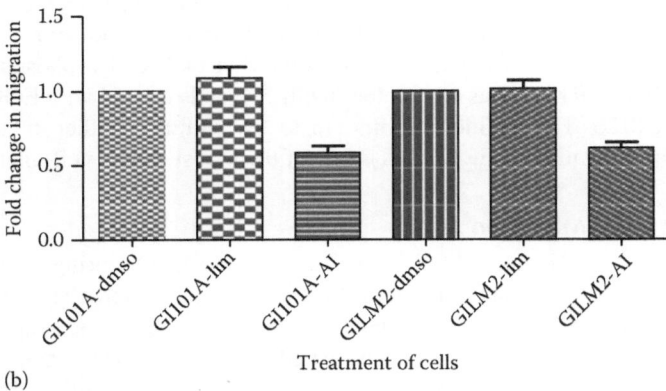

(b)

FIGURE 3.6   (a) Migration assays of Gl101A and GILM 2 cells in vitro. (b) Migration assays with normalization to DMSO control.

gene expression. As reported previously with several different human cancer cell lines (Tian et al., 2001), exposure to the limonin preparation has limited effect on growth of the breast cancer cells. The proliferation of one cell line (MDA-MB-231) was reduced by 25% by the highest dose tested (150 μM), while no reduction in growth of the other three cell lines was found. This unprecedented response may be due to physiologically higher concentrations of limonin toxic to this particular cell line. Our earlier studies demonstrated that limonin did not decrease the cell survival at the physiological concentrations of 1–5 μM concentrations (Somasundaram et al., 2012).

Also, treatment with limonin did not significantly alter the migration potential of two breast cancer cell lines in in vitro assays. It should be noted that the four cell lines in this study are ER negative and may be considered as models of triple-negative breast cancers. In the study by Tian group (Tian et al., 2001), growth inhibition of the MCF-7, ER-positive breast cancer cell line was reported; this may suggest that the effects of limonin may be restricted to a specific subtype of breast cancer cells. Consistent with the lack of effect on cell proliferation, the MAPK signaling pathway was not altered by short-term (6 h) exposure to limonin in the GILM2 and MDA-MB-231 cell lines. The impact on the NF-κB pathway was assessed, although in light of a previous report (Somasundaram et al., 2007), it might be worth noting that limonin induces phosphorylations of Ser 468 of NF-κB in MDA-MB-231 cells (Somasundaram et al., 2012).

Concerning the effect of limonin on ERK activation, our earlier reports (Somasundaram et al., 2012) suggested that limonin did not induce phosphorylations of ERK at 1, 5, and 10 µM concentrations. At the same time, limonin induces phosphorylations of Ser 468 of NF-κB in MDA-MB-231 cells. However, in the present study, in Figure 3.5, we observed that there was a slight increase in phosphorylations of ERK, which may be due to the inhibition of acid phosphatase enzyme at higher 100 µM concentrations in MDA-MB-231 cells (Li et al., 1998).

Also, the observed differential effect of a slight increase in ERK phosphorylations may be due the fact that the MDA-MB-231 cells primarily originated from breast cancer tissue, whereas the GILM2 is a breast cancer metastasis variant cell line of the lungs. This result again warrants a new future study to find the mechanism of the efficacy of drugs on cancerous tissues that originate from different organs of the body.

3.2.3.1   Gene Expression Profile for Metastasis-Associated Genes

The array chosen for the study had genes from several classes of protein factors involved in the process of metastasis. These included factors potentially regulating cell adhesion, extracellular matrix components and matrix-degrading proteases, cell cycle regulators, cell growth and proliferation, apoptosis, transcription factors and regulators, and other genes related to metastasis. Data from seven or eight PCR plates were combined for analysis (data of one plate from the GI101A group were lost from the analysis due to a technical failure in the PCR machine). This data and the full gene list are presented in Excel files, attached as Tables 3.A.1 through 3.A.3. In Table 3.A.1 are the results of comparing the expression data of GI101A and GILM2 cells treated with DMSO, the control condition for the experiment. This comparison was conducted to test whether the PCR array plates could detect differential gene expression between these two isogenic cell lines. The previous analysis done with these cell lines used a different system for analysis. In contrast, the array plates focused the study on the expression of genes specifically related to metastasis. Table 3.1 shows the genes that were found to be differentially expressed (greater than two-fold difference) with p values of 0.05. The genes include several matrix metalloproteases, including MMP7 that was previously reported as upregulated in GILM2 and TIMP2, an inhibitor of MMPs. The PCR array identified factors that may contribute to

TABLE 3.1

Genes Differentially Expressed by GILM2 Cells Relative to GI101A Cells

Gene Symbol	Description	Fold-Change	P Value
COL4A2	Collagen, type IV, alpha 2	0.052	0.025
CXCL12	Chemokine (C-X-C motif) ligand 12 (stromal-derived factor 1)	0.074	0.031
EPHB2	EPH receptor B2	9.79	0.017
ETV4	Ets variant gene 4 (E1A enhancer-binding protein, E1AF)	0.25	0.03
HGF	Hepatocyte growth factor	0.482	0.038
METAP2	Methionyl aminopeptidase 2	0.42	0.03
MMP11	Matrix metallopeptidase 11 (stromelysin 3)	0.354	0.001
MMP7	Matrix metallopeptidase 7 (matrilysin)	2.83	0.065
MMP9	Matrix metallopeptidase 9 (92 kDa type IV collagenase)	2.36	0.001
PLAUR	Plasminogen activator, urokinase receptor	0.423	0.034
TGFB1	Transforming growth factor, beta 1	0.211	0.021
TIMP2	TIMP metallopeptidase inhibitor 2	0.355	0.001

Note: Data from the PCR array analyses for DMSO-treated samples of GI101A and GILM2 cells, from seven replicate plates for GI101A, and from eight replicates for GILM2. Analysis of the data was performed using the web portal provided by SABiosciences Corp. The table consists of genes that were differentially expressed at least two-fold, with a p value < 0.05. The exception is MMP7, with $p = 0.065$, which reveals a trend to significantly higher expression in GILM2 cells. The p value calculations are based on Student's t-tests of the replicate $2^{-\Delta Ct}$ values for each gene.

tumor microenvironment modulation, HGF, TGF-β, and CXCL12/SDF1-α, with reduced expression detected in the metastatic variant cells.

The analysis of the comparison of DMSO- and limonin-treated GI101A cells is presented in Table 3.A.2. The expression of none of the 84 genes on the array was significantly changed by exposure to limonin at the 25 μM dose selected for the study. The expression of BRMS1 (breast cancer metastasis suppressor 1) was reduced by 4.16-fold, yet this difference was not statistically significant ($p = 0.6$) because of variability between samples; the standard deviation of Ct values for the limonin-treated samples is 3.5, while for the DMSO-treated samples the value is 0.49 (see Average Ct Table 3.A.2). Similarly for the GILM2 cells, limonin treatment had little effect on the expression of the genes in the PCR array. The only gene that was differentially expressed was integrin beta 3 with a 2.17-fold increase in limonin-treated cells ($p = 0.045$) (Table 3.A.3).

To examine whether limonin treatment can modulate the expression of genes for angiogenic factors, real-time PCR was performed using the same RNA samples used for the PCR arrays, plus an additional independently collected set of control and limonin-treated samples. Gene expression of the factors (VEGF-A, IL-8, CXCL1, and CXCL12) by limonin-treated cells is shown in Figures 3.7 and 3.8 and expressed as a percentage of the expression in DMSO-treated cells. The values from three independent experiments were averaged to derive the values shown in these figures.

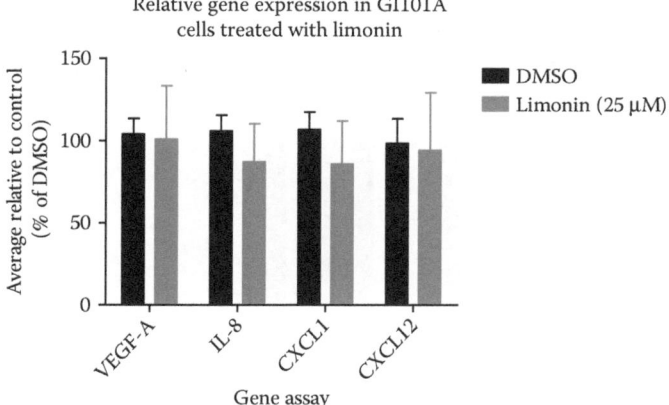

Bonferroni's multiple comparison test	Mean diff.	t	Significant? P < 0.05?	Summary	95% CI of diff
DMSO -vegf vs. lim -vegf f	3.312	0.2558	No	ns	−30.56 to 37.18
DMSO-il8 vs. lim il8 8	18.93	1.462	No	ns	−14.94 to 52.80
DMSO cxcl1 vs. lim cxcl1 1	21.28	1.643	No	ns	−12.59 to 55.15
DMSO cxcl12 vs. lim cxcl12 2	4.367	0.3372	No	ns	−29.50 to 38.24

FIGURE 3.7 Expression of VEGF-A, IL-8, CXCL1, and CXCL12 by DMSO- and limonin-treated GI101A cells, shown relative values from DMSO-treated cells: The data shown are averaged from the analysis of three independently collected samples. Analysis of variance, with Bonferroni's multiple comparison test, found no significant differences in expression following limonin treatment.

An analysis of variance with Bonferroni's multiple comparison test found no significant change in the expression of any of the factors in the limonin-treated cells in either cell lines. For GI101A cells, there was a modest reduction in IL-8 and CXCL1 expression following limonin treatment, but this result was variable and not consistently found in repeat experiments.

Gene expression levels of metastasis-associated and several angiogenic factors were compared in two breast cancer cell lines. These cancer cell lines have a common origin and differ in tumorigenic and metastatic ability in immunodeficient mice (Lev, Kiriakova, and Price, 2003). The cell lines showed significant differences in the expression of several of the 84 metastasis-associated genes surveyed in a focused PCR array format. However, no significant differences in gene expression were seen in limonin-treated GI101A cells compared with control samples, and the expression of one gene was found to be significantly altered by limonin treatment only in the GILM2 cell line (Figure 3.9). This was integrin beta 3 subunit; the protein product, as a heterodimer with αv, forms the vitronectin receptor, which is involved in interactions with the extracellular matrix. This is the first report that nutritionally important citrus-derived limonin activates integrin beta 3 subunit with respect to a specific cancer cell line. This integrin receptor has been shown to regulate various cellular processes involved in the regulation of tissue development, inflammation,

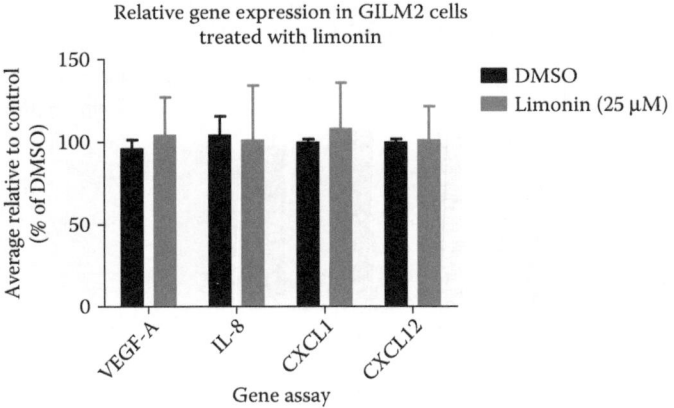

Bonferroni's multiple comparison test	Mean diff.	t	Significant? $P < 0.05$?	Summary	95% CI of diff
DMSO -vegf vs. lim -vegf f	−8.020	0.7251	No	ns	−37.02 to 20.98
DMSO-il8 vs. lim il8 8	2.775	0.2509	No	ns	−26.23 to 31.78
DMSO cxcl1 vs. lim cxcl1 1	−8.225	0.7436	No	ns	−37.23 to 20.78
DMSO cxcl12 vs. lim cxcl12 2	−1.452	0.1174	No	ns	−33.88 to 30.98

FIGURE 3.8 Expression of VEGF-A, IL-8, CXCL1, and CXCL12 by DMSO- and limonin-treated GILM2 cells, shown relative values from DMSO-treated cells: The data shown are averaged from the analysis of three independently collected samples. Analysis of variance, with Bonferroni's multiple comparison test, found no significant differences in expression following limonin treatment.

angiogenesis, metastasis, and apoptosis (Desgrosellier and Cheresh, 2010). Further investigation to confirm the limonin-induced increase in integrin beta 3 in the metastatic variant line will be done in future experiments.

Assessment of the effect of limonin on the expression of several angiogenic chemokines was prompted in part by the finding in the report of Vanamala (Vanamala et al., 2006) of reduced iNOS and COX-2 in rats fed with diets containing limonin. These are both mediators of chronic inflammation regulated through the NF-κB pathway; the same pathway is involved in modulating the expression of VEGF-A, IL-8, and other chemokines (Huang et al., 2000; Grivennikov and Karin, 2010). However, no significant changes in expression of these genes were seen in breast cancer cells exposed to limonin in vitro. One possibility for the lack of a measurable effect of limonin on the breast cancer cells is that the dose used for these in vitro experiments, notably the migration and gene expression experiments, was too low and that higher concentrations might have led to significant changes. In addition, the dose used (25 μM) was estimated to approach physiologically achievable doses to simulate what might occur in vivo.

In summary, this pilot study using several in vitro assays has failed to demonstrate any significant effect of limonin on estrogen-receptor negative human breast cancer cells.

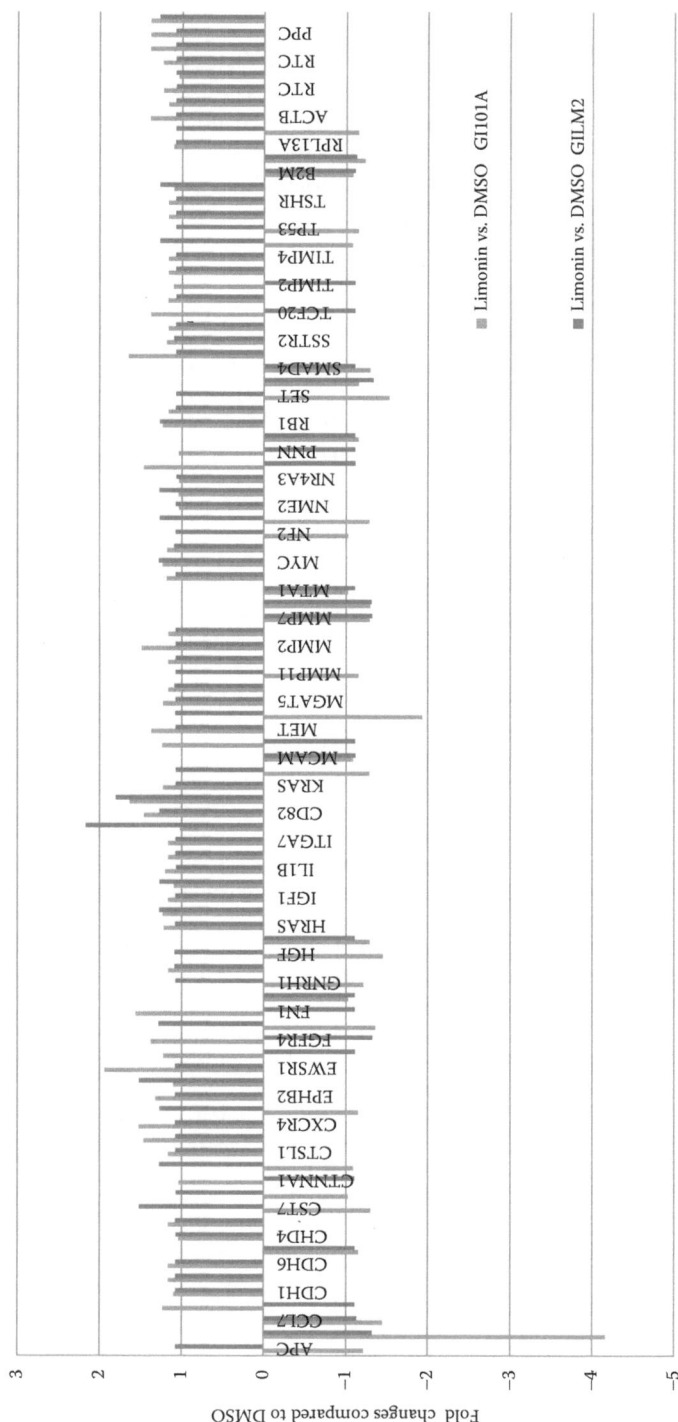

FIGURE 3.9 **(See color insert.)** Natural product citrus limonin interaction with metastazing genome of human breast cancer cells.

ACKNOWLEDGMENTS

We thank Dr. Daniel Jaeckle, University of Houston-Victoria (UHV), for reading and editing the manuscript with his critical comments. This project is funded by the USDA-CSREES # 2009-34402-19831 Designing Foods for Health through the Vegetable & Fruit Improvement Center and UHV Faculty Development Research Grant 2008. Part of this result was presented in the Cancer Genome Atlas conference, MD Anderson Cancer Center, Houston, Texas, April 2011.

REFERENCES

Apostolakis, S., E. G. Papadakis, E. Krambovitis, and D. A. Spandidos. 2006. Chemokines in vascular pathology (review). *Int J Mol Med* 17 (5):691–701.

Boomgaarden, I., S. Egert, G. Rimbach, S. Wolffram, M. J. Müller, and F. Döring. 2010. Quercetin supplementation and its effect on human monocyte gene expression profiles in vivo. *Br J Nutr* 104 (3):336–45.

Chelouche-Lev, D., H. M. Kluger, A. J. Berger, D. L. Rimm, and J. E. Price. 2004. alphaB-crystallin as a marker of lymph node involvement in breast carcinoma. *Cancer* 100 (12):2543–8.

Damås, J. K., C. Smith, E. Øie et al. 2007. Enhanced expression of the homeostatic chemokines CCL19 and CCL21 in clinical and experimental atherosclerosis: possible pathogenic role in plaque destabilization. *Arterioscler Thromb Vasc Biol* 27 (3):614–20.

Desgrosellier, J. S., and D. A. Cheresh. 2010. Integrins in cancer: biological implications and therapeutic opportunities. *Nat Rev Cancer* 10 (1):9–22.

Grivennikov, S. I., and M. Karin. 2010. Dangerous liaisons: STAT3 and NF-kappaB collaboration and crosstalk in cancer. *Cytokine Growth Factor Rev* 21 (1):11–9.

Hishikawa, K., T. Nakaki, and T. Fujita. 2005. Oral flavonoid supplementation attenuates atherosclerosis development in apolipoprotein E-deficient mice. *Arterioscler Thromb Vasc Biol* 25 (2):442–6.

Huang, S., J. B. Robinson, A. Deguzman, C. D. Bucana, and I. J. Fidler. 2000. Blockade of nuclear factor-kappaB signaling inhibits angiogenesis and tumorigenicity of human ovarian cancer cells by suppressing expression of vascular endothelial growth factor and interleukin 8. *Cancer Res* 60 (19):5334–9.

Kluger, H. M., D. Chelouche Lev, Y. Kluger et al. 2005. Using a xenograft model of human breast cancer metastasis to find genes associated with clinically aggressive disease. *Cancer Res* 65 (13):5578–87.

Lev, D. C., G. Kiriakova, and J. E. Price. 2003. Selection of more aggressive variants of the gI101A human breast cancer cell line: a model for analyzing the metastatic phenotype of breast cancer. *Clin Exp Metastasis* 20 (6):515–23.

Li, H., T. Miyahara, Y. Tezuka et al. 1998. The effect of Kampo formulae on bone resorption in vitro and in vivo. I. Active constituents of Tsu-kan-gan. *Biol Pharm Bull* 21 (12):1322–6.

Manach, C., C. Morand, A. Gil-Izquierdo, C. Bouteloup-Demange, and C. Rémésy. 2003. Bioavailability in humans of the flavanones hesperidin and narirutin after the ingestion of two doses of orange juice. *Eur J Clin Nutr* 57 (2):235–42.

Mandadi, K. K., G. K. Jayaprakasha, N. G. Bhat, and B. S. Patil. 2007. Red Mexican grapefruit: a novel source for bioactive limonoids and their antioxidant activity. *Z Naturforsch C* 62 (3–4):179–88.

Milenkovic, D., C. Deval, C. Dubray, A. Mazur, and C. Morand. 2011. Hesperidin displays relevant role in the nutrigenomic effect of orange juice on blood leukocytes in human volunteers: a randomized controlled cross-over study. *PLoS One* 6 (11):e26669.

Mink, P. J., C. G. Scrafford, L. M. Barraj et al. 2007. Flavonoid intake and cardiovascular disease mortality: a prospective study in postmenopausal women. *Am J Clin Nutr* 85 (3):895–909.

Onishi, K., A. Higuchi, T. Sakura, N. Masuyama, and Y. Gotoh. (2007) The PI3K-Akt pathway promotes microtubule stabilization in migrating fibroblasts. Genes cells. 12(4):535–546.

Patil, J. R., K. N. Chidambara Murthy, G. K. Jayaprakasha, M. B. Chetti, and B. S. Patil. 2009. Bioactive compounds from Mexican lime (Citrus aurantifolia) juice induce apoptosis in human pancreatic cells. *J Agric Food Chem* 57 (22):10933–42.

Singh, S., and B. B. Aggarwal. 1995. Activation of transcription factor NF-kappa B is suppressed by curcumin (diferuloylmethane) [corrected]. *J Biol Chem* 270 (42):24995–5000.

Skubitz, K. M., and A. P. Skubitz. 2008. Interdependency of CEACAM-1, -3, -6, and -8 induced human neutrophil adhesion to endothelial cells. *J Transl Med* 6:78.

Somasundaram, S., K. Pearce, R. Gunasekera, G. K. Jayaprakasha, and B. Patil. 2007. Differential phosphorylations of NFkB and cell growth of MDA-MB231 human breast cancer cell line by limonins. *Acta Hort* 841:55–57.

Somasundaram, S., J. Price, K. Pearce, R. Shuck, G. K. Jayaprakasha, and B. Patil. 2012. Citrus limonin lacks the antichemotherapeutic effect in human models of breast cancer. *J Nutrigenet Nutrigenomics* 5 (2):106–14.

Sørlie, T., C. M. Perou, R. Tibshirani et al. 2001. Gene expression patterns of breast carcinomas distinguish tumor subclasses with clinical implications. *Proc Natl Acad Sci U S A* 98 (19):10869–74.

Tian, Q., E. G. Miller, H. Ahmad, L. Tang, and B. S. Patil. 2001. Differential inhibition of human cancer cell proliferation by citrus limonoids. *Nutr Cancer* 40 (2):180–4.

Urquidi, V., and S. Goodison. 2007. Genomic signatures of breast cancer metastasis. *Cytogenet Genome Res* 118 (2–4):116–29.

van de Vijver, M. J., Y. D. He, L. J. van't Veer et al. 2002. A gene-expression signature as a predictor of survival in breast cancer. *N Engl J Med* 347 (25):1999–2009.

Vanamala, J., T. Leonardi, B. S. Patil et al. 2006. Suppression of colon carcinogenesis by bioactive compounds in grapefruit. *Carcinogenesis* 27 (6):1257–65.

Wang, P., Z. Liu, C. Wu, B. Zhu, Y. Wang, and H. Xu. 2008. Evaluation of CD86/CD28 and CD40/CD154 pathways in regulating monocyte-derived CD80 expression during their interaction with allogeneic endothelium. *Transplant Proc* 40 (8):2729–33.

Weber, C., A. Schober, and A. Zernecke. 2004. Chemokines: key regulators of mononuclear cell recruitment in atherosclerotic vascular disease. *Arterioscler Thromb Vasc Biol* 24 (11):1997–2008.

Weigelt, B., A. M. Glas, L. F. Wessels, A. T. Witteveen, J. L. Peterse, and L. J. van't Veer. 2003. Gene expression profiles of primary breast tumors maintained in distant metastases. *Proc Natl Acad Sci U S A* 100 (26):15901–5.

Weigelt, B., L. F. Wessels, A. J. Bosma et al. 2005. No common denominator for breast cancer lymph node metastasis. *Br J Cancer* 93 (8):924–32.

Wong, B. W., D. Wong, and B. M. McManus. 2002. Characterization of fractalkine (CX3CL1) and CX3CR1 in human coronary arteries with native atherosclerosis, diabetes mellitus, and transplant vascular disease. *Cardiovasc Pathol* 11 (6):332–8.

APPENDIX 3.A

TABLE 3.A.1
Supplementary Data Set 1: Gl101A versus GILM2

Reaction Well Numbers	Symbol	AVG Ct		Standard Deviation	
		GILM2	Gl101A	GILM2	Gl101A
A01	APC	30.15	29.41	0.962714	0.991158
A02	BRMS1	27.65	27.65	0.493986	0.496631
A03	CCL7	34.15	33.14	0.969349	2.053225
A04	CD44	22.9	23.65	0.007447	0.500895
A05	CDH1	22.14	22.14	0.947162	0.956375
A06	CDH11	35	35	0	0
A07	CDH6	35	35	0	0
A08	CDKN2A	25.14	25.14	0.9552	0.957152
A09	CHD4	25.65	25.4	0.495262	0.573869
A10	COL4A2	28.9	24.9	0.808901	1.152151
A11	CST7	33.4	32.64	0.584602	0.495852
A12	CTBP1	25.89	25.65	0.004598	0.495275
B01	CTNNA1	22.4	22.39	0.576702	0.990099
B02	CTSK	29.64	29.89	0.502503	0.815121
B03	CTSL1	35	35	0	0
B04	CXCL12	31.4	27.9	0.573776	1.154036
B05	CXCR4	34.98	33.94	0.042404	1.223682
B06	DENR	25.38	25.14	0.578828	0.957744
B07	EPHB2	30.9	34.45	0.010395	0.639704
B08	ETV4	33.92	32.17	0.819091	0.496338
B09	EWSR1	26.41	26.65	0.576758	0.494331
B10	FAT1	23.39	23.64	0.575105	0.492132
B11	FGFR4	29.41	30.15	0.574641	0.500026
B12	FLT4	28.9	28.9	0.807668	0.813632
C01	FN1	30.16	30.66	0.95664	1.497826
C02	FXYD5	23.15	22.65	0.497729	0.493761
C03	GNRH1	29.67	29.41	0.497807	0.998634
C04	KISS1R	35	35	0	0
C05	HGF	34.95	34.14	0.061185	0.966305
C06	HPSE	27.66	27.65	0.953709	0.501015
C07	HRAS	25.4	25.65	0.582746	0.504159
C08	HTATIP2	24.4	24.64	0.577151	0.498065
C09	IGF1	35	35	0	0
C10	IL18	26.39	26.15	0.580538	0.9566
C11	IL1B	34.14	34.64	0.503214	0.495016
C12	IL8RB	35	35	0	0
D01	ITGA7	35	35	0	0
D02	ITGB3	34.49	34.49	0.593199	1.018969
D03	CD82	24.9	24.9	0.814682	0.004094

(Continued)

TABLE 3.A.1 (*Continued*)
Supplementary Data Set 1: GI101A versus GILM2

Reaction Well Numbers	Symbol	AVG Ct GILM2	AVG Ct GI101A	Standard Deviation GILM2	Standard Deviation GI101A
D04	KISS1	31.4	31.39	0.572013	0.576443
D05	KRAS	26.4	26.66	0.579418	0.49527
D06	RPSA	21.91	21.66	0.816562	0.503505
D07	MCAM	32.65	32.9	0.503964	0.008238
D08	MDM2	25.65	25.65	0.504544	0.499548
D09	MET	28.39	29.14	0.578001	0.502641
D10	METAP2	29.41	28.41	1.28623	0.580431
D11	MGAT5	27.9	27.65	0.002141	0.495844
D12	MMP10	34.95	34.97	0.055584	0.053797
E01	MMP11	28.4	27.15	0.579436	0.96162
E02	MMP13	35	35	0	0
E03	MMP2	35	33.93	0	1.189853
E04	MMP3	35	35	0	0
E05	MMP7	27.9	29.65	1.824871	0.502597
E06	MMP9	29.15	30.64	0.953857	0.499101
E07	MTA1	24.89	24.65	0.00601	0.494153
E08	MTSS1	35	35	0	0
E09	MYC	28.16	27.66	0.502976	0.493401
E10	MYCL1	34.93	34.95	0.044457	0.053487
E11	NF2	26.15	25.65	0.514292	0.492368
E12	NME1	22.9	22.65	0.813972	0.500624
F01	NME2	20.4	20.4	0.579924	0.995187
F02	NME4	24.4	24.41	0.575544	0.568684
F03	NR4A3	34.17	34.42	0.963502	0.587235
F04	PLAUR	28.89	27.9	0.006416	0.817957
F05	PNN	24.16	24.41	0.957269	1.00246
F06	PTEN	24.4	24.14	0.57114	0.951568
F07	RB1	26.39	25.65	0.577349	0.500513
F08	RORB	35	35	0	0
F09	SET	22.9	22.4	0.816076	1.001703
F10	SMAD2	27.17	27.17	0.962635	0.953804
F11	SMAD4	25.92	25.67	0.815797	0.492449
F12	SRC	28.91	28.42	0.002247	0.580295
G01	SSTR2	34.93	35	0.050877	0
G02	SYK	35	35	0	0
G03	TCF20	27.39	27.14	0.580792	0.953867
G04	TGFB1	29.89	27.89	0.003946	0.003952
G05	TIMP2	26.4	25.16	0.572172	0.962398
G06	TIMP3	35	35	0	0
G07	TIMP4	35	35	0	0
G08	TNFSF10	25.4	24.91	0.574728	1.415196

(*Continued*)

TABLE 3.A.1 (*Continued*)
Supplementary Data Set 1: GI101A versus GILM2

Reaction Well Numbers	Symbol	AVG Ct		Standard Deviation	
		GILM2	GI101A	GILM2	GI101A
G09	TP53	24.15	24.15	0.494725	0.497906
G10	TRPM1	35	35	0	0
G11	TSHR	35	35	0	0
G12	VEGFA	26.14	26.15	0.500965	0.49413
H01	B2M	21.16	21.92	0.958827	0.817822
H02	HPRT1	23.15	23.4	0.95725	1.000037
H03	RPL13A	22.92	22.16	0.826767	0.955796
H04	GAPDH	17.66	18.16	0.955022	0.954924
H05	ACTB	18.66	19.16	0.957214	1.501203
H06	HGDC	35	35	0	0
H07	RTC	25.64	25.64	0.956478	0.954929
H08	RTC	25.64	25.39	0.956379	0.579588
H09	RTC	25.64	25.64	0.955838	0.954755
H10	PPC	21.9	22.15	0.014938	0.49731
H11	PPC	21.9	22.15	0.013657	0.498308
H12	PPC	22.14	22.15	0.500502	0.499466

TABLE 3.A.2
Supplementary Data Set 2: GI101A DMSO versus Limonin

Reaction Well Numbers	Symbol	AVG Ct		Standard Deviation	
		Limonin	DMSO	Limonin	DMSO
A01	APC	29.9	29.41	0.008165	0.991158
A02	BRMS1	29.93	27.65	3.505882	0.496631
A03	CCL7	33.89	33.14	1.004898	2.053225
A04	CD44	23.57	23.65	0.578169	0.500895
A05	CDH1	22.22	22.14	0.570599	0.956375
A06	CDH11	35	35	0	0
A07	CDH6	35	35	0	0
A08	CDKN2A	25.56	25.14	0.585181	0.957152
A09	CHD4	25.56	25.4	0.575308	0.573869
A10	COL4A2	24.89	24.9	0.003661	1.152151
A11	CST7	33.23	32.64	0.579469	0.495852
A12	CTBP1	25.9	25.65	0.006434	0.495275
B01	CTNNA1	22.57	22.39	0.576676	0.990099
B02	CTSK	30.23	29.89	0.58012	0.815121
B03	CTSL1	35	35	0	0
B04	CXCL12	27.56	27.9	1.154329	1.154036
B05	CXCR4	33.55	33.94	1.145575	1.223682

(*Continued*)

TABLE 3.A.2 (*Continued*)
Supplementary Data Set 2: GI101A DMSO versus Limonin

Reaction		AVG Ct		Standard Deviation	
Well Numbers	Symbol	Limonin	DMSO	Limonin	DMSO
B06	DENR	25.55	25.14	0.581615	0.957744
B07	EPHB2	34.26	34.45	1.180921	0.639704
B08	ETV4	32.26	32.17	0.57304	0.496338
B09	EWSR1	25.91	26.65	0.003732	0.494331
B10	FAT1	23.57	23.64	0.571181	0.492132
B11	FGFR4	29.91	30.15	0.005046	0.500026
B12	FLT4	29.56	28.9	0.578167	0.813632
C01	FN1	30.24	30.66	0.581504	1.497826
C02	FXYD5	22.9	22.65	0.00301	0.493761
C03	GNRH1	29.91	29.41	1.000874	0.998634
C04	KISS1R	35	35	0	0
C05	HGF	34.9	34.14	0.006973	0.966305
C06	HPSE	28.23	27.65	1.159281	0.501015
C07	HRAS	25.58	25.65	0.572687	0.504159
C08	HTATIP2	24.56	24.64	0.582295	0.498065
C09	IGF1	35	35	0	0
C10	IL18	26.23	26.15	0.578362	0.9566
C11	IL1B	34.6	34.64	0.614412	0.495016
C12	IL8RB	35	35	0	0
D01	ITGA7	35	35	0	0
D02	ITGB3	34.67	34.48	0.572878	1.011625
D03	CD82	24.56	24.9	0.570998	0.004094
D04	KISS1	30.9	31.39	0.0116	0.576443
D05	KRAS	26.58	26.66	0.586308	0.49527
D06	RPSA	22.24	21.66	0.573702	0.503505
D07	MCAM	33.23	32.9	0.5746	0.008238
D08	MDM2	25.55	25.65	0.579131	0.499548
D09	MET	28.9	29.14	0.995247	0.502641
D10	METAP2	29.58	28.41	0.580017	0.580431
D11	MGAT5	27.57	27.65	0.579417	0.495844
D12	MMP10	34.97	34.97	0.051122	0.053797
E01	MMP11	27.57	27.15	0.571424	0.96162
E02	MMP13	35	35	0	0
E03	MMP2	33.57	33.93	1.156313	1.189853
E04	MMP3	35	35	0	0
E05	MMP7	30.23	29.65	0.581267	0.502597
E06	MMP9	31.23	30.64	1.156108	0.499101
E07	MTA1	24.89	24.65	0.003762	0.494153
E08	MTSS1	34.97	35	0.051052	0
E09	MYC	27.58	27.66	0.57991	0.493401
E10	MYCL1	34.93	34.95	0.057609	0.053487

(Continued)

TABLE 3.A.2 (*Continued*)
Supplementary Data Set 2: GI101A DMSO versus Limonin

Reaction Well Numbers	Symbol	AVG Ct		Standard Deviation	
		Limonin	DMSO	Limonin	DMSO
E11	NF2	25.9	25.65	0.009025	0.492368
E12	NME1	23.23	22.65	0.572684	0.500624
F01	NME2	20.57	20.4	0.57731	0.995187
F02	NME4	24.57	24.41	0.579943	0.568684
F03	NR4A3	34.59	34.42	0.584644	0.587235
F04	PLAUR	27.56	27.9	0.564178	0.817957
F05	PNN	24.58	24.41	0.57953	1.00246
F06	PTEN	24.56	24.14	0.575958	0.951568
F07	RB1	25.56	25.65	0.575442	0.500513
F08	RORB	35	35	0	0
F09	SET	23.23	22.4	0.576958	1.001703
F10	SMAD2	27.59	27.17	0.585899	0.953804
F11	SMAD4	26.26	25.67	0.573215	0.492449
F12	SRC	27.92	28.42	0.004832	0.580295
G01	SSTR2	34.97	35	0.054944	0
G02	SYK	35	35	0	0
G03	TCF20	26.9	27.14	0.006532	0.953867
G04	TGFB1	27.89	27.89	0.00562	0.003952
G05	TIMP2	25.24	25.16	0.56683	0.962398
G06	TIMP3	35	35	0	0
G07	TIMP4	35	35	0	0
G08	TNFSF10	25.23	24.91	0.574991	1.415196
G09	TP53	24.57	24.15	0.579773	0.497906
G10	TRPM1	35	35	0	0
G11	TSHR	35	35	0	0
G12	VEGFA	26.24	26.15	1.145809	0.49413
H01	B2M	22.25	21.92	0.571456	0.817822
H02	HPRT1	23.91	23.4	1.006603	1.000037
H03	RPL13A	22.24	22.16	0.578797	0.955796
H04	GAPDH	18.57	18.16	0.576928	0.954924
H05	ACTB	18.91	19.16	0.990524	1.501203
H06	HGDC	35	35	0	0
H07	RTC	25.56	25.64	0.577756	0.954929
H08	RTC	25.55	25.39	0.577084	0.579588
H09	RTC	25.56	25.64	0.577255	0.954755
H10	PPC	21.89	22.15	0.002186	0.49731
H11	PPC	21.9	22.15	0.004344	0.498308
H12	PPC	21.9	22.15	0.006549	0.499466

TABLE 3.A.3
Supplementary Data Set 3: Comparison of Limonin-Treated GI101A and GILM2 Cells on Metastasis Genome Expression

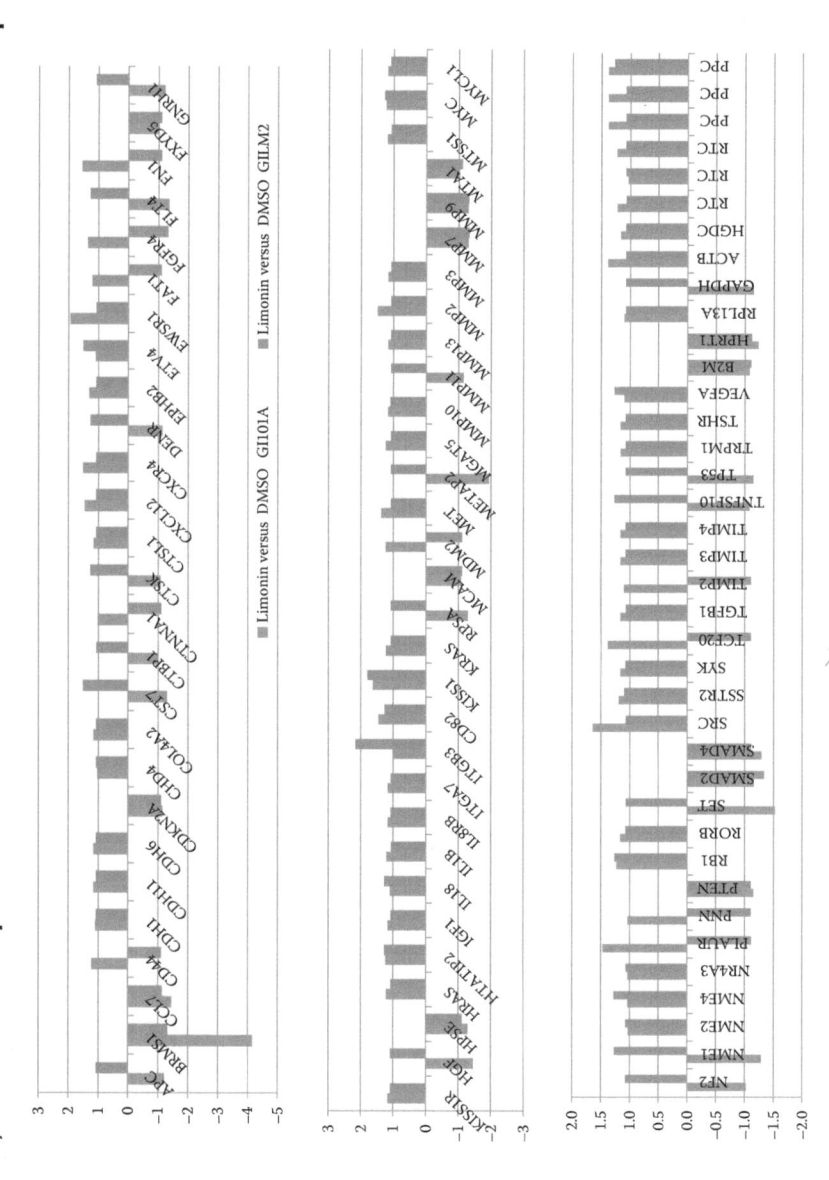

4 Combinatorial Effect of Natural Products and Chemotherapy Interactions in Cancer

Siva G. Somasundaram, Janet Price,
G. K. Jayaprakasha, Richard S.
Gunasekera, and Bhimanagouda S. Patil

CONTENTS

4.1   INTRODUCTION

We have previously demonstrated that citrus-derived bioactive molecules can inhibit the proliferation of colon, pancreatic, and breast cancer cells (Tian et al. 2001; Jayaprakasha et al. 2008; Patil et al. 2009; Chidambara Murthy et al. 2011). Among the different citrus limonoids examined, obacunone was the most potent inhibitor of cancer cell proliferation (Poulose, Harris, and Patil 2005; Somasundaram et al. 2007). In addition to their independent antiproliferative ability, limonoids such as limonin and its glucoside have an additive colon cancer inhibition effect when combined with curcumin (Chidambara Murthy, Jayaprakasha, and Patil 2013). This inhibition activity results from their ability to induce programmed cell death in cancer cells (Tian et al. 2001; Poulose, Harris, and Patil 2005) and activate phase II enzymes in cancer-induced animals (Tanaka et al. 2001). Research at our laboratory also demonstrated the anti-aromatase activity of obacunone, along with other limonoids and glucosides, in human breast cancer cells (Kim, Jayaprakasha, and Patil 2013). Obacunone and obacunone glucoside (OG) occur in very low concentrations in citrus fruits, from 7.2 to 60 ppm in different parts of the fruits (Ozaki et al. 1991, 1995). In a rat model of azoxymethane-induced colon cancer (aberrant crypt foci formation), limonoids were shown to suppress proliferation and enhance apoptosis by lowering levels of cyclooxygenase-2 (COX-2) and inducible nitric oxide synthase (iNOS) induced by carcinogen (Vanamala et al. 2006), potentially via the NF-κB pathway. This NF-κB signaling pathway can regulate many genes that are important for tumor progression and metastasis (Singh and Aggarwal 1995), as discussed in Chapter 3, but whether limonin can affect the expression of metastasis-associated genes has not been previously reported. In addition, the spice-derived agent curcumin has been shown to modulate a variety of signaling pathways in cancer cells and is reported to have anti-cancer activities (Miquel et al. 2002; Bharti, Donato, and Aggarwal 2003; Aggarwal et al. 2004). This study was designed to test whether exposure to a combination of limonin and curcumin can modulate the expression of genes associated with angiogenesis and metastasis in human breast cancer cells and impact the malignant potential of human breast cancer cells using a xenograft model. Evidently, our previous studies proved that curcumin inhibits the traditional chemotherapeutic activities both in vitro and in vivo (Somasundaram et al. 2002). In this chapter, we present breast cancer and prostate cancer studies with and without chemotherapies.

4.2   INTERACTION OF NATURAL PRODUCTS CITRUS LIMONIN AND CURCUMIN WITH BREAST CANCER GENES

The present study used a human breast cancer line that is tumorigenic and metastatic in immunodeficient mice. It has been previously reported that the differential gene expression patterns in this variant compared with the poorly metastatic isogenic variant (Kluger et al. 2005). The metastatic cell line displays high gene activity in the mitogen-activated protein kinase (MAPK) and NF-κB signaling pathways, both of which have been shown to be modulated by the citrus bioactive compound limonin and/or the spice derivative curcumin. The experiments determine whether these agents alone or in combination can inhibit or reduce gene expression of metastasis-associated genes and/or factors that promote tumor angiogenesis. The studies were performed on cells in tissue culture and on xenograft tumors in immunodeficient mice. RNA samples isolated from the breast cancer cells were exposed either in vitro or in vivo to limonin and/or curcumin and were used for polymerase chain reaction (PCR) analyses to establish whether these agents can modify angiogenesis and metastasis-associated gene expression.

4.2.1   METHODS AND MATERIALS

4.2.1.1   Cell Line

The GILM2 variant of the GI101A human breast cancer cell line, derived as reported previously (Lev, Kiriakova, and Price 2003), was maintained in a monolayer culture in Dulbecco's minimum essential medium (4 mM glucose), supplemented with 10% fetal calf serum, L-glutamine, and penicillin–streptomycin. The cells were tested for and found free of *Mycoplasma* sp. infection.

4.2.1.2   In Vitro Responses to Limonin

Tumor cells were plated in 96-well culture plates at an initial density of 1×10^4 cells per well and allowed to attach for 24 h. The culture medium was then changed to medium with different concentrations of limonin from a stock solution prepared in dimethyl sulfoxide (DMSO) at a concentration of 100 mM and with curcumin, which was also dissolved in DMSO. The control condition was medium with an equivalent concentration of DMSO at the highest concentration of limonin or curcumin. The cells were incubated for up to 72 h; the relative cell numbers were assessed using MTT. The conversion of MTT to formazan in metabolically viable cells was monitored with an MR-5000 microtiter plate reader set to read at 570 nm.

4.2.1.3   Gene Expression Assays

The GILM2 cells were plated into 100 mm plates in DMEM with 10% fetal calf serum. The following day when the cultures were of approximately 80% confluence, the culture medium was replaced with medium supplemented with 0.05% DMSO or 25 µM limonin, 25 µM curcumin, or a combination of limonin and curcumin. The cultures were incubated for a further 24 h, and then cells were collected for the isolation of total RNA using the ArrayGrade Total RNA Isolation Kit (SABiosciences Corp., Frederick, MD). Real-time PCR was performed for selected genes using the total RNA samples reverse-transcribed with random primers from the High Capacity cDNA Archive Kit (Applied Biosystems, Foster City, CA). cDNA was amplified in duplicate samples using Predeveloped TaqMan Assay Reagents for VEGF, IL-8, MMP-9, CRYAB, and 18S following the manufacturer's recommended amplification procedure. Results were recorded as mean threshold cycle (Ct), and relative expression was determined using the comparative Ct method. The ΔCt was calculated as the difference between the average Ct value of the endogenous control (18S) and the average Ct value of the test gene. To compare the relative amount of target gene expression in different samples, human placenta RNA was used as a calibrator.

The $\Delta\Delta$Ct was determined by subtracting the ΔCt of the calibrator from the ΔCt of the test sample. Relative expression of the target gene was calculated by the formula $2^{-\Delta\Delta Ct}$, which is the amount of gene product normalized to the endogenous control and relative to the calibrator sample.

4.2.1.4 In Vivo Experiment

Human GILM2 breast cancer cells were injected into the mammary fat pad of 40 female nude mice, injecting 5×10^6 cells per mouse. On day 26 after tumor injection, the mice were randomized into four groups; tumor volumes at this time were approximately 100 mm^3 calculated from the formula: volume $= xy^2/2$, where x is the largest diameter and y is the smallest diameter of the tumor, measured using calipers. The four groups received treatment as follows:

1. Control: intraperitoneal (IP) injection of phosphate buffered saline (PBS) and daily gavage of corn oil, as control for curcumin feeding.
2. IP injection of limonin (10 μM, diluted in PBS from a 100 μM preparation in DMSO, injection of 0.1 mL) at 2-day intervals for a total of nine injections and daily gavage of corn oil.
3. Daily feeding by gavage of curcumin, 20 mg/day in 0.1 mL corn oil, five times per week, for a total of 14 doses.
4. Combination of curcumin and limonin at same doses/schedule as single-agent groups.

Tumor sizes were measured twice weekly during the course of treatment. One day after the final injection of limonin (day 43 after tumor injection), the animals were euthanized; the tumors were removed, weighed, and sample fixed or frozen for histology, immunohistochemistry, and RNA extraction. Snap frozen pieces of tumor were ground in a pestle and mortar, and total RNA was isolated using Qiagen RNeasy reagents.

4.2.2 Results

4.2.2.1 In Vitro Sensitivity of Breast Cancer Cells to Limonin

The MTT assay was used to evaluate the response of GILM2 cells to limonin and curcumin, using a single concentration of limonin (25 μM) and a range of concentrations of curcumin (0–50 μM). Previous studies with limonin alone had found no or modest effect of the agent alone on the growth and viability of the cells. The data shown in Figure 4.1 show the sensitivity of the cells to curcumin and that the addition of limonin to the culture medium has no effect on the response of cells to curcumin.

4.2.2.2 In Vivo Response of GILM2 Tumors to Chronic Treatment with Limonin and Curcumin

The injection of GILM2 cells produced progressively growing tumors in 36 mice (90%). These mice were randomized into four treatment groups with treatment started on day 26 after tumor cell injection. The treatments were apparently well

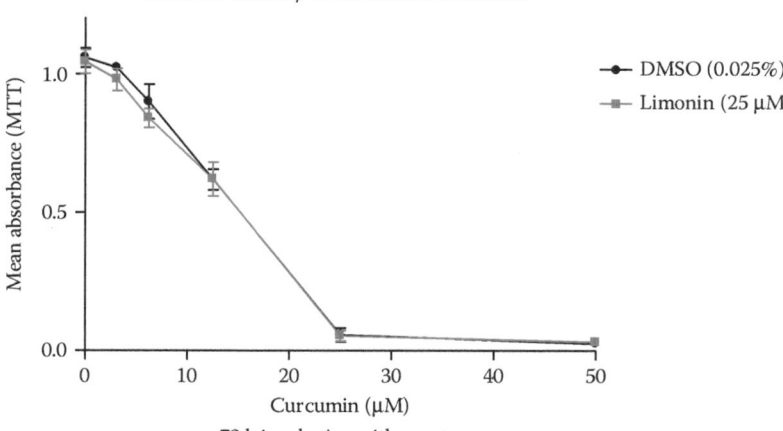

FIGURE 4.1   (See color insert.) Response of GILM2 cells to limonin and curcumin treatment in vitro: The results of MTT assays for relative cell numbers in cultures of GILM2 cells treated with limonin and a range of concentrations of curcumin. The MTT assay was performed 72 h after addition of the agents.

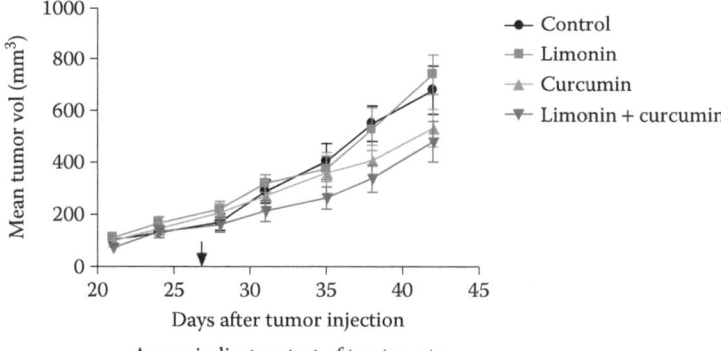

FIGURE 4.2   (See color insert.) Response of human breast cancer xenograft tumors to limonin and curcumin treatments: Human GILM2 breast cancer cells were injected in the mammary fat pads of immunodeficient mice. When tumors were of approximately 100 mm³ volume, mice were treated daily with curcumin (oral gavage, 20 mg/day in corn oil) or by IP injection with a limonin (10 μM, 0.1 mL at 2-day intervals). Control group mice were fed corn oil and injected with 0.1 mL PBS. Tumors were measured twice weekly, and treatment continued for 20 days.

tolerated with no obvious weight loss or adverse effects noted over the course of the experiment. Figure 4.2 shows the growth curves of the tumors of mice in different treatment groups, with the tumors in mice receiving curcumin or curcumin plus limonin showing a delay in growth compared with the tumors in mice in the control group or the tumors in mice receiving limonin injections. The weights of tumors at the end of the experiment are shown in Figure 4.3; the tumors from mice receiving

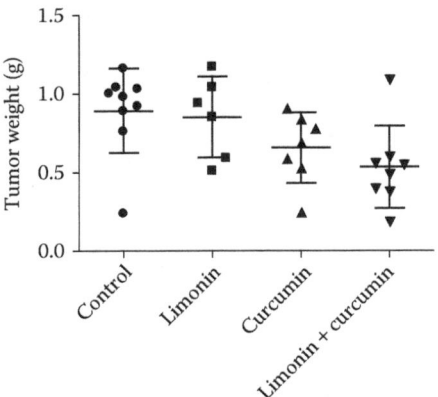

Tukey's multiple comparison test	Mean diff.	q	Significant? P < 0.05?	Summary	95% CI of diff.
Control vs. limonin	0.04111	0.4338	No	ns	−0.3269 to 0.4091
Control vs. curcumin	0.2354	2.597	No	ns	−0.1164 to 0.5872
Control vs. limonin + curcumin	0.3586	4.104	Yes	*	0.01937 to 0.6979
Limonin vs. curcumin	0.1943	1.942	No	ns	−0.1941 to 0.5827
Limonin vs. limonin + curcumin	0.3175	3.269	No	ns	−0.05955 to 0.6945
Curcumin vs. limonin + curcumin	0.1232	1.324	No	ns	−0.2381 to 0.4845

FIGURE 4.3 Weights of tumors from mice treated with curcumin and limonin: The combination of agents resulted in a significant reduction in tumor growth compared with tumors from animals in the control group.

the two agents were significantly smaller than those of animals in other treatment groups ($p = 0.0309$, by ANOVA [analysis of variance]).

Immunohistochemical staining of sections of frozen sections of the tumors was performed to detect CD-31-positive tumor-associated blood vessels. Sections from five tumors of each treatment group were stained and scored for the presence of stained vessels in 10 representative fields per slide. The results, shown in Figure 4.4, suggest that the treatment with curcumin and limonin had no effect on the formation of tumor-associated blood vessels in the GILM2 xenograft tumors. Sections of formalin-fixed and paraffin-embedded tumors were stained with hematoxylin and eosin and used to score for the proportions of cells undergoing mitosis (scoring mitotic figures). The mitotic index of tumors from mice treated with curcumin was significantly reduced compared with tumors from mice in the control group or tumors from mice treated with limonin consistent with the observed reduction in tumor growth (Figure 4.5).

The GILM2 xenograft tumors are metastatic, seeding metastases to the lungs of immunodeficient mice. Sections of lungs from each of the mice in the experiment were scored for micrometastases, and the numbers are shown in Figure 4.6. Although there was the most variation in numbers of metastases in lungs of curcumin-treated mice, overall there was no reduction in the incidence or numbers of metastases in any of mice in the treatment groups.

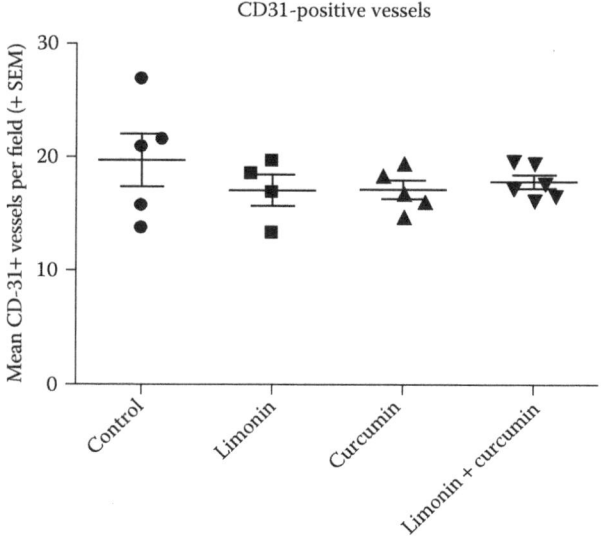

FIGURE 4.4 Analysis of tumor sections for CD-31 positivity did not show any difference in tumor-associated blood vessels, suggesting that the treatments did not alter tumor angiogenesis.

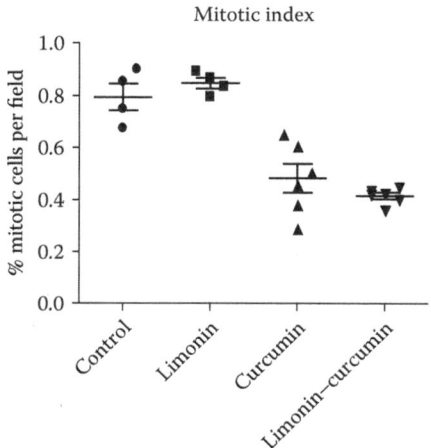

Tukey's multiple comparison test	Mean diff.	q	Significant? P < 0.05?	Summary	95% CI of diff.
Control vs. limonin	−0.05225	1.146	No	ns	−0.2367 to 0.1322
Control vs. curcumin	0.3110	7.471	Yes	***	0.1426 to 0.4794
Control vs. limonin–curcumin	0.3778	9.077	Yes	***	0.2094 to 0.5462
Limonin vs. curcumin	0.3633	8.727	Yes	***	0.1948 to 0.5317
Limonin vs. limonin–curcumin	0.4301	10.33	Yes	***	0.2617 to 0.5985
Curcumin vs. limonin–curcumin	0.06683	1.795	No	ns	−0.08380 to 0.2175

FIGURE 4.5 Scoring for mitotic figures in tissue sections showed that curcumin treatment reduced tumor proliferation.

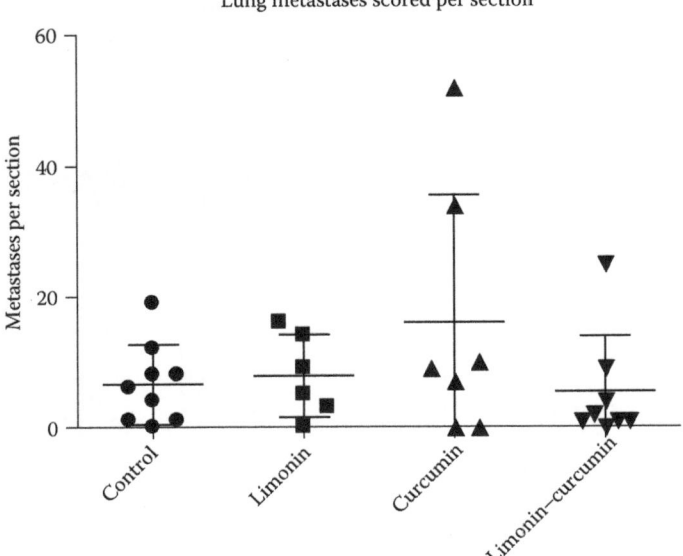

FIGURE 4.6 Lung metastases in GILM2 tumor-bearing mice: Hematoxylin and eosin–stained sections of the lungs of mice were scored for the presence of micrometastases. The incidence and numbers of metastases per section were not different for samples from different treatment groups.

4.2.2.3 Measurement of Angiogenic Factor Gene Expression in GILM2 Breast Cancer Cells

RNA isolated from GILM2 cells cultured for 24 h in the presence of curcumin and limonin was used for PCR measurements of gene expression. As the goal of the study was to analyze effects on angiogenesis, three genes linked to the development of tumor vasculature, VEGF, IL-8, and MMP-9, were investigated. All these genes have also been shown to be regulated through pathways sensitive to inhibition by curcumin, notably the NF-κB pathway (Singh and Aggarwal 1995). The results shown in Figure 4.7 show a trend toward the inhibition of VEGF and MMP-9 expression in cells exposed to curcumin. However, the only significant difference found in this experiment was a significant increase in IL-8 expression in cells exposed to curcumin. The data are from two replicate experiments; possibly more study with additional time points or concentrations is needed to understand this response. Total RNA was isolated from five tumors of each treatment group and used for reverse-transcription than real-time PCR measurement of expression of VEGF, IL-8, MMP-9, and CRYAB, relative to 18S. CRYAB is a small heat shock protein, which we previously reported as increased in the GILM2 variant and linked to shorter survival in breast cancer patients (Chelouche-Lev et al. 2004). Figure 4.8 shows that the relative expression levels of these genes in GILM2 xenograft tumors are not altered by treatment with curcumin, limonin, or the combination of these agents.

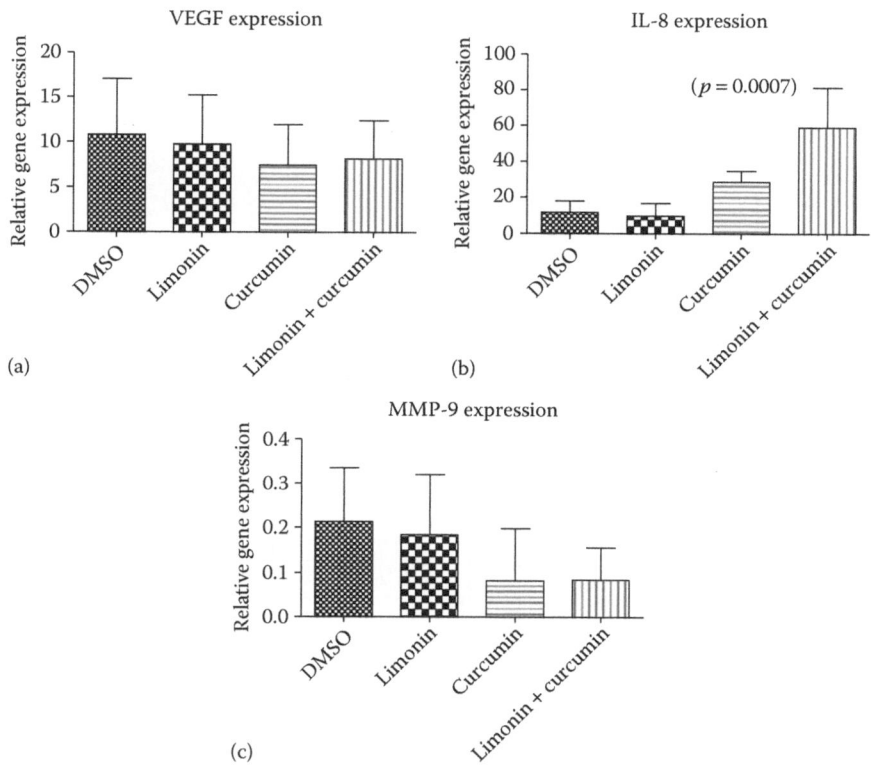

(a)

(b)

(c)

FIGURE 4.7   Regulation of gene expression in GILM2 cells treated in vitro with limonin and curcumin: RNA isolated from GILM2 cells exposed for 24 h to limonin or curcumin was used for RT-PCR measurements of the expression of (a) VEGF, (b) IL-8, and (c) MMP-9. Relative gene expression of the gene of interest, relative to 18S, was calculated using the ΔΔCt method (see Section 4.2.1). ANOVA was used to compare expression between treatments; the increase in IL-8 expression in cells treated with limonin and curcumin was significantly different from other groups.

4.2.3 DISCUSSION

Citrus-derived limonin is reported to have various activities that may affect the survival and proliferation of cancer cells, although the molecular mechanisms involved have not been fully identified (Tian et al. 2001; Vanamala et al. 2006; Patil et al. 2009). The assessment of limonin's effect on the expression of several angiogenic chemokines was prompted in part by the finding in the report of Vanamala et al. (2006) that reduced iNOS and COX-2 in rats fed with diets containing limonin. Both these are mediators of chronic inflammation regulated through the NF-κB pathway; the same pathway is involved in modulating the expression of VEGF-A, IL-8, and other chemokines (Huang et al. 2000; Grivennikov and Karin 2010). However, no significant changes in expression of these genes were seen in breast cancer cells exposed to limonin in vitro or in tumors of chronically treated mice.

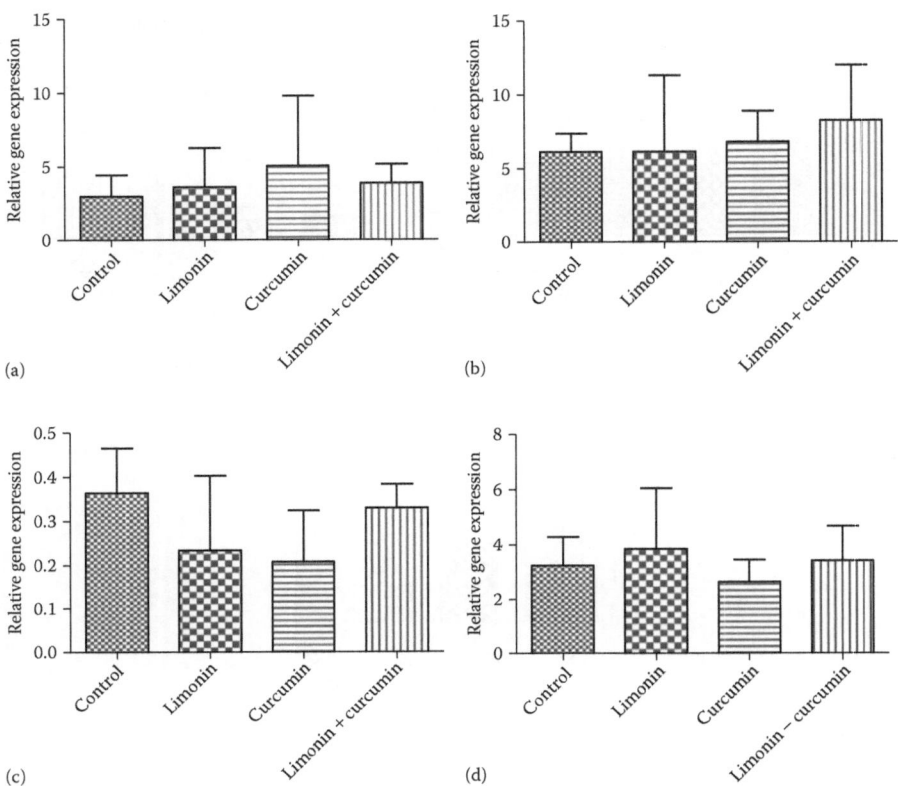

(a)   (b)   (c)   (d)

FIGURE 4.8 Measurements of gene expression using RNA isolated from tumors of mice treated with limonin and curcumin: RNA isolated from GILM2 tumors from mice treated with limonin or curcumin was used for RT-PCR measurements of the expression of (a) VEGF, (b) IL-8, (c) MMP-9, and (d) CRYAB. Relative gene expression of the gene of interest, relative to 18S, was calculated using the $\Delta\Delta$Ct method (see Section 4.2.1). Values shown are means from four or five tumor samples per treatment group. There were no significant differences between the groups for any of the genes tested.

The responses of cancer cells to curcumin have been well documented, and this agent is reported to affect many signaling pathways resulting in changes in gene expression. Curcumin has been proposed as a chemopreventive agent and for use in combination with chemotherapy. A previous pre-clinical study showed that the combination of curcumin and paclitaxel reduced the outgrowth of lung metastases of a human xenograft tumor model (Aggarwal et al. 2005). The present study showed that curcumin treatment with and without limonin slowed the growth of the GILM2 xenograft tumors. However, the attenuation of tumor growth was not apparently a function of an effect on angiogenic factor production by the tumors. There were no reduction in the numbers of tumor-associated blood vessels (identified by CD-31 immunoreactivity) and no difference in the expression of angiogenic factors in the tumors.

A key limitation of some pre-clinical models testing the effects of curcumin is in estimating an appropriate dose for an experiment, which would be achievable and tolerated in possible clinical studies. The protocol used of daily oral gavage of 20 mg per mouse (approximating to 0.8 g/kg) approaches that used by other investigators (Kunnumakkara et al. 2007) and did control the growth of GILM2 xenografts, although the response was relatively modest. Whether a larger dose or use of a formulation with improved bioavailability would produce better tumor growth control remains to be investigated. The addition of limonin did not have a clear effect, either negative or positive, on that of curcumin alone. The results overall did not provide support for the notion that the combination of these agents may affect the expression of angiogenic factors by breast cancer cells and have a potential application for anti-angiogenic therapies. In Section 4.3, we extend the interaction of natural products for the activities of prostate cancer genes.

4.3 INTERACTION OF NATURAL PRODUCTS CAPSAICIN, LUTEIN, AND CURCUMIN WITH AND WITHOUT TRADITIONAL CHEMOTHERAPIES ON PROSTATE CANCER GENES

Epidemiological studies reveal that high intake of green vegetables and fruits such as tomato and avocado and spices such as turmeric might prevent the risk of prostate cancer (Potter and Steinmetz 1996). Increased expression of prostate-specific antigen (PSA) in prostate cancer cells is a useful diagnostic marker. Also, PSA synthesis is regulated by NF-κB (Sun et al. 2010). COX-2 activity in prostate carcinogenesis is well established (Ferruelo et al. 2014). Camptothecin, a standard chemotherapy, is found to increase c-Jun N-terminal kinase (JNK) activity and inhibit the translocation of NF-κB (Somasundaram et al. 2002). It is evident that prolonged use of chemotherapy leads to cancer cells becoming resistant to that particular chemotherapeutic drug. Recent studies on phytochemicals such as curcumin and capsaicin sensitize the chemoresistance cancer cells for apoptosis (Saha et al. 2012). Curcumin being a COX-2 inhibitor plays a critical role in inhibiting NF-κB and JNK and preventing cancer cell growth. We assessed the efficacy of lutein (Bioflavone-alpha), capsaicin, and curcumin in different combinations, as well as in combination with camptothecin in human prostate cancer cell lines (LNCaP and PC3), measuring cell growth and corresponding levels of phospho JNK as a signaling marker gene activity.

4.3.1 RESULTS

Figure 4.8 shows that the cell growth inhibitory activities of curcumin on traditional chemotherapy camptothecin induced cell growth inhibition at 3 h. Curcumin at 1–10 μM concentrations inhibits the cell growth. However, at 10 μM concentration with camptothecin, curcumin did not inhibit the cell growth and exhibits its anti-apototic activities just like in breast cancer, as we reported earlier (Somasundaram et al. 2002). Figure 4.9 indicates the cell growth inhibition of capsaicin, lutein curcumin, and camptothecin on LNCaP cells at 4 h. In this, capsaicin is effective at 5 and 10 μM

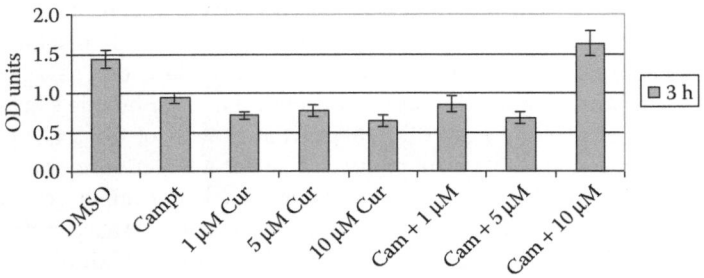

FIGURE 4.9 The effect of curcumin and camptothecin on LNCaP cell growth inhibition.

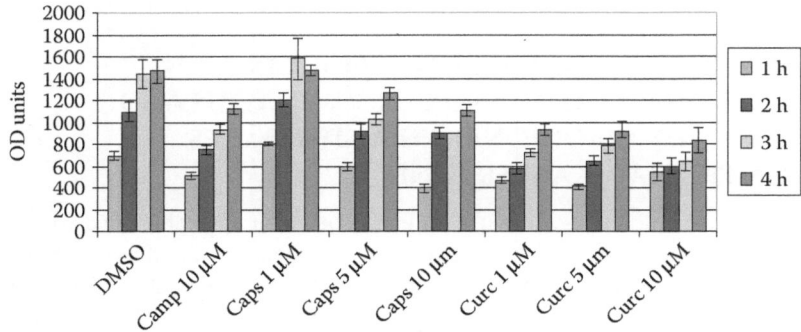

FIGURE 4.10 The effect of camptothecin, capsaicin, and curcumin on LNCaP cells.

concentration and curcumin is inhibiting from 1, 5, and 10 μM concentrations to 50% of the cell growth.

Figure 4.10 shows the effect of capsaicin is more pronounced to exhibit its cell death activities at 5, 10, 15, and 20 μM concentrations ($p < 0.05$). Figure 4.11 indicates that the cell growth inhibitory effect of bioflavone α (lutein) on PC3 cells is highly significant at 5 and 10 μM concentrations. Figure 4.12 shows the JNK activities with the bioflavone α (lutein) and capsaicin. This JNK gene is a signaling molecule that increases during apoptosis cell death. On the contrary, curcumin does decrease the JNK activities in breast cancer cells as well as in prostate cancer cells. Both lutein and capsaicin significantly increase this enzyme activity. Figure 4.13 indicates that capsaicin and bioflavone α together inhibit PC3 cell growth inhibition possibly through the signaling molecule of JNK gene activation. In this study, bioflavone α at 10 μM concentration compared with capsaicin at 5 μM concentration inhibits the cell growth more than 50%. To further confirm these findings, the JNK activity again proved that camptothecin at 2 μM concentration can inhibit more than 50% cell growth compared with bioflavone α at 10 μM concentration in PC3 cells (Figure 4.14). Thus, the requirement of traditional chemotherapy is so minimal with the natural product bioflavone α and also with the combination of two different natural products such

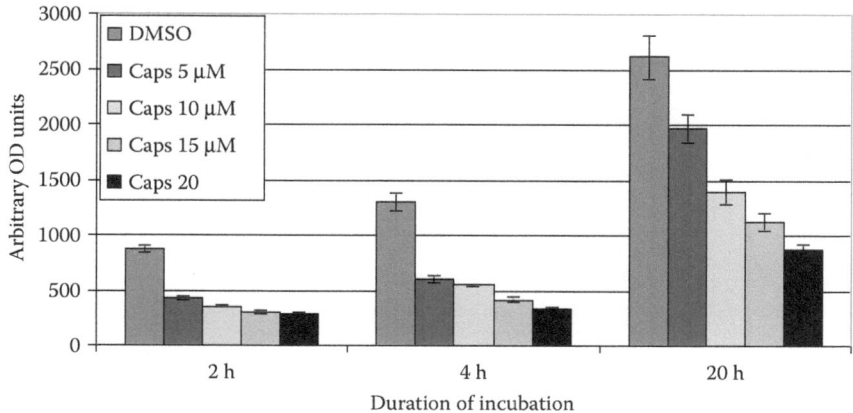

FIGURE 4.11   The effect of capsaicin on PC3 cell growth inhibition.

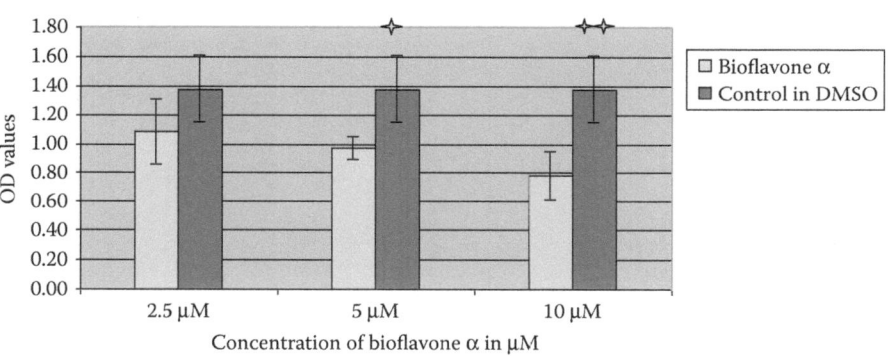

FIGURE 4.12   The cell growth inhibitory effect of bioflavone α (lutein) on PC3 cells.

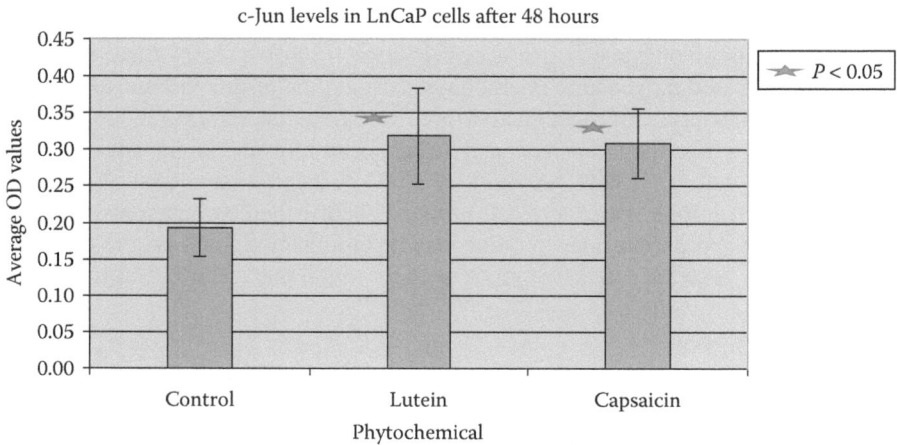

FIGURE 4.13   The JNK activities with the lutein and capsaicin on PC3 cells.

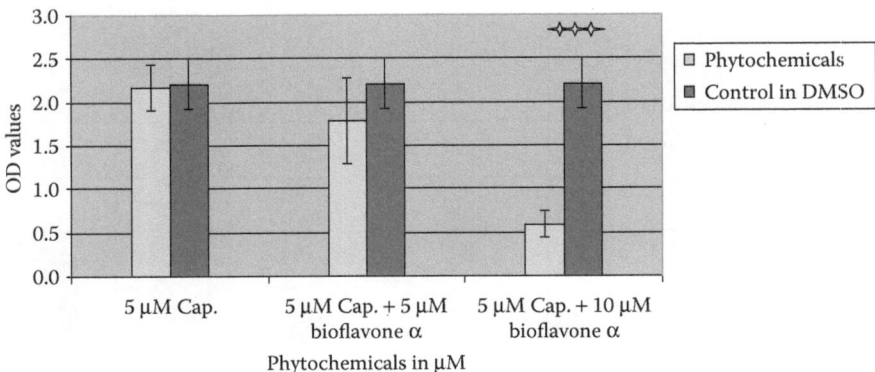

FIGURE 4.14 The effect of capsaicin and bioflavone α together on PC3 cell growth inhibition.

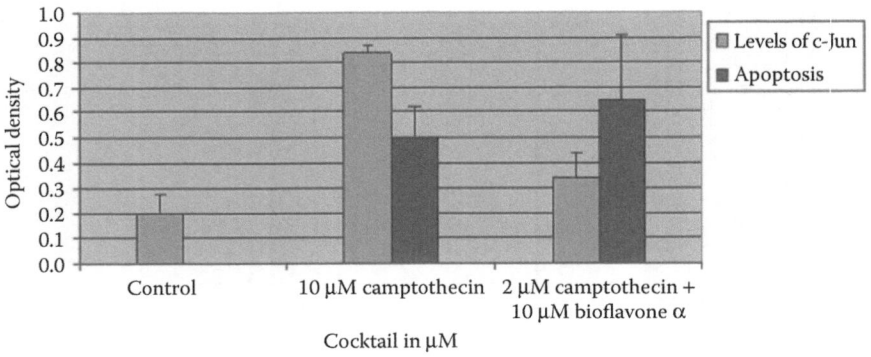

FIGURE 4.15 Combinatorial effects of camptothecin and bioflavone α on JNK activities and apoptosis in PC3 cells.

as capsaicin and bioflavone α interacting with each other to bring forth new drug formulations based on genomic interactions. Further, the results in Figure 4.15 indicate the combinatorial anticancer activities of curcumin, capsaicin, and bioflavone α in PC3 cells. In this study, capsaicin at 5 μM concentration and bioflavone α at 5 μM concentration compared with curcumin at 1 μM concentration exhibit an excellent combination to inhibit the cancer cell growth. Finally, the combinations of natural products at an appropriate concentration may be promising a significant role in altering the cancer genomic profile without a need for traditional chemotherapy.

4.3.2 DISCUSSION

Curcumin, the major component of the spice turmeric, is used as coloring, flavoring, and additive agents in many foods. It also exerts anti-inflammatory and chemopreventive activities from time immemorial. However, this agent is also a known inhibitor of reactive oxygen species (ROS) and the JNK pathway (Somasundaram et al. 2002). Many chemotherapeutic drugs generate ROS and activate JNK in the course of inducing apoptosis. Hence, the probability of curcumin might antagonize the antitumor activity of chemotherapeutic drugs when it is supplemented with diet (Somasundaram et al. 2002). Studies in tissue culture revealed that curcumin-inhibited camptothecin induced apoptosis only at a higher concentration (10 µM) in LNCaP human prostate cancer cell line, in time, and not in a concentrated-dependent manner (Figures 4.9 and 4.10). At lower concentrations (1 and 5 µM), curcumin exhibits neither decrease nor increase in the response to induce apoptosis with camptothecin. The modulatory activity of curcumin may be due to the property of antitumor activity, and at a higher concentration, 10 µM, it exerts its anti-oxidative property, thereby inhibiting the chemotherapeutic response in LNCaP prostate cell line.

A 48 h incubation of prostate cancer cells each at 5 µM concentration or camptothecin at 10 µM concentration induced cell growth inhibition depicting altered expressions of these molecular parameters. Interestingly, the modulation of these cellular and molecular parameters was more pronounced in cells treated with lower doses (of lutein and capsaicin, lutein and curcumin, or lutein and camptothecin at 5 µM concentration) rather than cells treated with higher doses of each individual agent. In conclusion, the present study demonstrates for the first time that a combination of lower doses of lutein (Bioflavone-alpha), capsaicin, and curcumin with camptothecin effectively modulates these cellular and molecular genomic mechanisms that are found in prostate cancer (Figure 4.16). This promises the use of low doses of dietary supplements in combination and phytochemotherapy with the assurance of minimal side effects. In Chapter 5, we focus on the effect of various natural products that interact with microbes genomes.

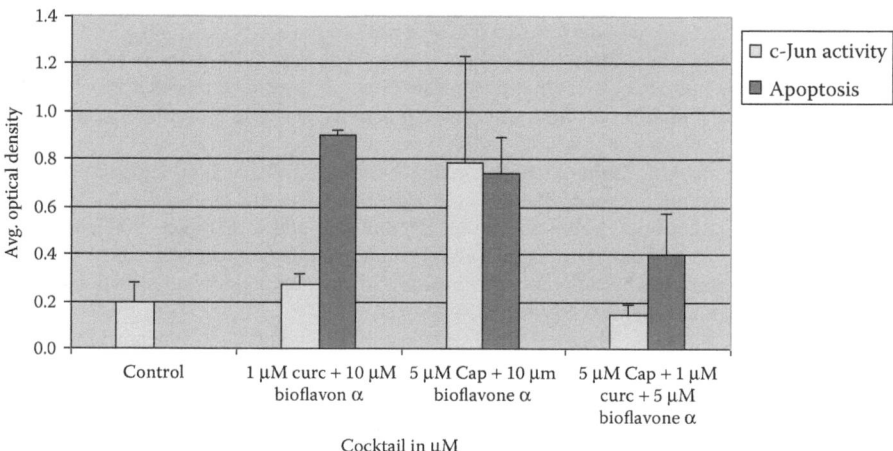

FIGURE 4.16   (See color insert.) The combinatorial anticancer activities of curcumin, capsaicin, and bioflavone α in PC3 cells.

ACKNOWLEDGMENTS

This project is funded by the USDA-CSREES # 2009-34402-19831 Designing Foods for Health through the Vegetable & Fruit Improvement Center and UHV Faculty Development Research Grant 2009. Part of this result was presented in the Cancer Prevention Research Institute of Texas (CPRIT), Austin, April 2010. The copy right permission also acknowledged for reproducing the Figures in this chapter from our earlier Publication during editing of the book. Reproduced from Richard S. Gunasekera, Siva G. Somasundaram, Morgan Mitchell, Desiree Arrambide and Homer S. Black, 2014, Lutein inhibits growth of human prostate cancer cells and potentiates capsaicin, curcumin, and the traditional chemotherapy agent, camptothecin, Current Topics in Phytochemistry, Vol. 12, 1–9.

REFERENCES

Aggarwal, B. B., S. Shishodia, Y. Takada et al. 2005. Curcumin suppresses the paclitaxel-induced nuclear factor-kappaB pathway in breast cancer cells and inhibits lung metastasis of human breast cancer in nude mice. *Clin Cancer Res* 11 (20):7490–8.

Aggarwal, S., Y. Takada, S. Singh, J. N. Myers, and B. B. Aggarwal. 2004. Inhibition of growth and survival of human head and neck squamous cell carcinoma cells by curcumin via modulation of nuclear factor-kappaB signaling. *Int J Cancer* 111 (5):679–92.

Bharti, A. C., N. Donato, and B. B. Aggarwal. 2003. Curcumin (diferuloylmethane) inhibits constitutive and IL-6-inducible STAT3 phosphorylation in human multiple myeloma cells. *J Immunol* 171 (7):3863–71.

Chelouche-Lev, D., H. M. Kluger, A. J. Berger, D. L. Rimm, and J. E. Price. 2004. AlphaB-crystallin as a marker of lymph node involvement in breast carcinoma. *Cancer* 100 (12):2543–8.

Chidambara Murthy, K. N., G. K. Jayaprakasha, and B. S. Patil. 2013. Citrus limonoids and curcumin additively inhibit human colon cancer cells. *Food Funct* 4 (5):803–810.

Chidambara Murthy, K. N., G. K Jayaprakasha, V. Kumar, K. S. Rathore, and B. S. Patil. 2011. Citrus limonin and its glucoside inhibit colon adenocarcinoma cell proliferation through apoptosis. *J. Agric. Food Chem.* 59 (6):2314–23.

Ferruelo, A., M. M. de Las Heras, C. Redondo, F. Ramón de Fata, I. Romero, and J. C. Angulo. 2014. Wine polyphenols exert antineoplasic effect on androgen resistant PC-3 cell line through the inhibition of the transcriptional activity of COX-2 promoter mediated by NF-kβ. *Actas Urol Esp.* 38(7):429–437.

Grivennikov, S. I., and M. Karin. 2010. Dangerous liaisons: STAT3 and NF-kappaB collaboration and crosstalk in cancer. *Cytokine Growth Factor Rev* 21 (1):11–19.

Huang, S., J. B. Robinson, A. Deguzman, C. D. Bucana, and I. J. Fidler. 2000. Blockade of nuclear factor-kappaB signaling inhibits angiogenesis and tumorigenicity of human ovarian cancer cells by suppressing expression of vascular endothelial growth factor and interleukin 8. *Cancer Res* 60 (19):5334–9.

Jayaprakasha, G. K., K. K. Mandadi, S. M. Poulose, Y. Jadegoud, G. A. Nagana Gowda, and B. S. Patil. 2008. Novel triterpenoid from Citrus aurantium L. possesses chemopreventive properties against human colon cancer cells. *Bioorg. Med. Chem.* 16 (11):5939–51.

Kim, J., G. K. Jayaprakasha, and B. S. Patil. 2013. Limonoids and their anti-proliferative and anti-aromatase properties in human breast cancer cells. *Food Funct* 4 (2):258–65.

Kluger, H. M., D. Chelouche Lev, Y. Kluger et al. 2005. Using a xenograft model of human breast cancer metastasis to find genes associated with clinically aggressive disease. *Cancer Res* 65 (13):5578–87.

Kunnumakkara, A. B., S. Guha, S. Krishnan, P. Diagaradjane, J. Gelovani, and B. B. Aggarwal. 2007. Curcumin potentiates antitumor activity of gemcitabine in an orthotopic model of pancreatic cancer through suppression of proliferation, angiogenesis, and inhibition of nuclear factor-kappaB-regulated gene products. *Cancer Res* 67 (8):3853–61.

Lev, D. C., G. Kiriakova, and J. E. Price. 2003. Selection of more aggressive variants of the gI101A human breast cancer cell line: a model for analyzing the metastatic phenotype of breast cancer. *Clin Exp Metastasis* 20 (6):515–23.

Miquel, J., A. Bernd, J. M. Sempere, J. Díaz-Alperi, and A. Ramírez. 2002. The curcuma antioxidants: pharmacological effects and prospects for future clinical use. A review. *Arch Gerontol Geriatr* 34 (1):37–46.

Ozaki, Y., S. Ayano, N. Inaba, M. Miyake, M. A. Berhow, and S. Hasegawa. 1995. Limonoid glucosides in fruit, juice and processing by-products of Satsuma Mandarin (Chus unshiu Marcov.). *J Food Sci* 60 (1):186–9.

Ozaki, Y., C. H. Fong, Z. Herman et al. 1991. Limonoid glucosides in citrus seeds. *Agric Biol Chem* 55 (1):137–41.

Patil, J. R., K. N. Chidambara Murthy, G. K. Jayaprakasha, M. B. Chetti, and B. S. Patil. 2009. Bioactive compounds from Mexican lime (Citrus aurantifolia) juice induce apoptosis in human pancreatic cells. *J Agric Food Chem* 57 (22):10933–42.

Potter, J. D., and K. Steinmetz. 1996. Vegetables, fruit and phytoestrogens as preventive agents. *IARC Sci Publ* 139:61–90.

Poulose, S. M., E. D. Harris, and B. S. Patil. 2005. Citrus limonoids induce apoptosis in human neuroblastoma cells and have radical scavenging activity. *J Nutr* 135 (4):870–7.

Saha, S., A. Adhikary, P. Bhattacharyya, T. Das, and G. Sa. 2012. Death by design: where curcumin sensitizes drug-resistant tumours. *Anticancer Res* 32 (7):2567–84.

Singh, S., and B. B. Aggarwal. 1995. Activation of transcription factor NF-kappa B is suppressed by curcumin (diferuloylmethane) [corrected]. *J Biol Chem* 270 (42):24995–5000.

Somasundaram, S., N. A. Edmund, D. T. Moore, G. W. Small, Y. Y. Shi, and R. Z. Orlowski. 2002. Dietary curcumin inhibits chemotherapy-induced apoptosis in models of human breast cancer. *Cancer Res* 62 (13):3868–75.

Somasundaram, S., K. Pearce, R. Gunasekera, G. K. Jayaprakasha, and B. Patil. 2007. Differential phosphorylations of NFkB and cell growth of MDA-MB 231 human breast cancer cell line by limonins. *Acta Hort* 841:55–57.

Sun, H. Z., T. W. Yang, W. J. Zang, and S. F. Wu. 2010. Dehydroepiandrosterone-induced proliferation of prostatic epithelial cell is mediated by NFKB via PI3K/AKT signaling pathway. *J Endocrinol* 204 (3):311–18.

Tanaka, T., M. Maeda, H. Kohno et al. 2001. Inhibition of azoxymethane-induced colon carcinogenesis in male F344 rats by the citrus limonoids obacunone and limonin. *Carcinogenesis* 22 (1):193–8.

Tian, Q., E. G. Miller, H. Ahmad, L. Tang, and B. S. Patil. 2001. Differential inhibition of human cancer cell proliferation by Citrus Limonoids. *Nutr Cancer* 40 (2):180–4.

Vanamala, J., T. Leonardi, B. S. Patil et al. 2006. Suppression of colon carcinogenesis by bioactive compounds in grapefruit. *Carcinogenesis* 27 (6):1257–65.

5 Interactions of Natural Products with Microbial Genomes

Chandra Somasundaram

CONTENTS

5.1 INTRODUCTION

It is estimated that humans harbor 10^{13}–10^{14} microbial cells. These cells mainly grow on the epithelial surface of the body. They present as a complex microbial community. Prevalent area includes skin, mouth, vagina, and gastrointestinal tract. Some of the microorganisms can grow in even acidic environments such as in the stomach where the pH is 2. This complex microbial community is generally called *microbiota* that consists of thousands of different microbial members. They include bacteria, archaea, fungi, viruses, and eukaryotic microorganisms present in specific epithelial tissues. We focus on the microbiota present in gastrointestinal tracts that influence the nutrition, metabolism, physiology, and immune function of the host. They participate in breaking down of complex carbohydrates that are present in regular diets and help the host. Similarly, this microbiota has the property to protect the host from pathogenic microbial organisms as well by competing for resources or even inhibiting the growth of the pathogen.

The following list provides the species of microorganisms that are present in various sites of gastrointestinal tract (Sartor, 2008).

1. Stomach harbors 0–10^2 organisms such as *Lactobacillus*, *Candida*, *Streptococcus*, *Helicobacter pylori*, and *Peptostreptococcus*.
2. Duodenum harbors 10^2 organisms such as *Streptococcus* and *Lactobacillus*.
3. Jejunum harbors 10^2 organisms such as *Streptococcus* and *Lactobacillus*.
4. Proximal ileum harbors 10^3 organisms such as *Streptococcus* and *Lactobacillus*.

5. Distal ileum harbors 10^7–10^8 organisms such as *Clostridium*, *Streptococcus*, bacteroids, *Actinomyciae*, and *Corynebacteria*.
6. Colon harbors 10^{11}–10^{12} organisms such as bacteroids, *Clostridium* groups IV and XIV, *Bifidobacterium*, and Enterobacteriaceae.

The recent investigations indicate that the distal gut microbiota of adult is different from adolescent children (Agans et al., 2011). Also, the human gut microbial gene catalog indicates (using the fecal sample cohort study) that over 99% of the genes are bacterial and each individual sharing 160 species (Qin et al., 2010). Both commensals and pathogens are included in the gut. It is interesting to understand the immunity against pathogens but not to commensal oraganisms. The mechanism is not fully understood yet. There is evidence that commensals may participate in the immune system, namely in T cells and B cells, generation, and function (Sathyabama, Khan, and Agrewala, 2014).

It has been mentioned in Chapters 2 through 4 that natural products have a significant impact on human genomes. Now the same natural products interact with the microorganisms that are present in the gut, alter the pharmacokinetics of the natural products metabolites availability and therapeutic activity, and participate in the biotransformations of the natural products. There are challenges in administering the natural products to increase the bioavailability for a better therapeutic activity without major side effects. The report on this topic further confirms that antibiotic administration reduces the biotransformation of orally administered drugs by gut microbiota resulting in a decrease in drug absorption and an increase in the pharmacokinetic drug interaction (Yoo et al., 2014).

At the same time, the natural products should cooperatively act with the symbiotic microorganisms to achieve the goal to heal. Instead, some natural products interact with healthy gut microbiota and destroy them and induce adverse side effects. The drug–microbiota interactions also pose major threat when we have a diet with the natural products that have the significant biological effects. As it has been mentioned earlier, the microbiota of adults is different from adolescent children; it is wise enough to choose the appropriate diet (natural products) for the drug. Similarly, it has been reported that the people living in various geographical regions have different microbiota in their gut and experience different intestinal absorption and permeability (Menzies et al., 1999). In addition, we have to consider race, gender, age, and genetic makeup of the individuals for harboring microbial populations in their gut. It is very important to note that the endotoxin produced by gut microbial community may even ignite an intense inflammatory response in the host organism that includes the formation of *p*-cresol and indoxyl sulfate during protein fermentation by the microbes. It is reported (Ramezani and Raj, 2014) that these endotoxins might breach the gut barriers that lead toxins to make entry into the systemic circulations to induce cardiovascular diseases, uremic toxicity, and septicemia. To avoid this unwanted microbiota activity, the use of probiotic therapy could help to reestablish the intestinal symbiosis by either neutralizing the bacterial endotoxins or removing the toxins, thereby maintaining the metabolically balanced intestinal microbiota. Thus, the role of natural products should maintain the gut symbiosis and inhibit selected pathogenic organisms.

Sometimes, drug therapies, such as chemotherapy, radiation therapy, or any other therapies that are administered during the gastrointestinal disorders, may pose a threat to the microbiota community that lives in the gut. After considering all these facts, the following sections are discussed with various natural products interacting with different microorganisms to prevent the growth both in vitro culture and in vivo models. For example, microbial communities help to transform quercetin to enhance the bioavailability, and in the colon, they participate in degrading the unabsorbed metabolites of quercetin (Graf et al., 2005).

5.2 DIFFERENT NATURAL PRODUCTS INTERACTING WITH SINGLE MICROORGANISMS

The natural products inhibit the growth of the microorganisms, and there is not much information on specific genes that are involved in the cell death of the organisms. However, certain natural products such as curcumin, a strong antioxidant, can modulate the genes involved in survival pathways during environmental oxidative stress.

There are three natural products, thymol, citral, and linalyl acetate, that interact with *Acenetobacter calcoaceticus* bacterium (Figure 5.1). This organism is present as a normal flora of skin and throat of humans. It may emerge as an opportunistic pathogen and has 3674 genes. These three natural products can inhibit the growth of *Acenetobacter* (Piccaglia et al., 1993; Dorman and Deans, 2000).

Figure 5.2 discusses the effects of three natural products, geraniol, linalyl acetate, and alpha-terpineol, on the microorganism, *Alcaligenes faecalis* (Dorman and Deans, 2000).

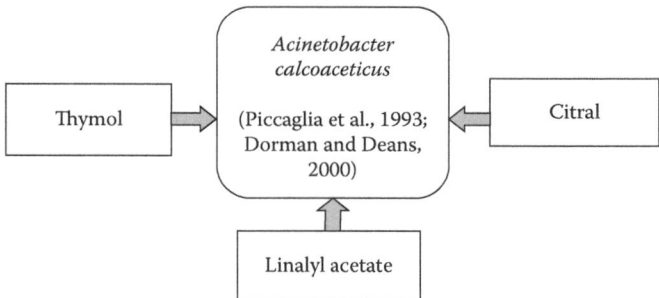

FIGURE 5.1 Natural products interacting with microorganism *Acinetobacter calcoaceticus*.

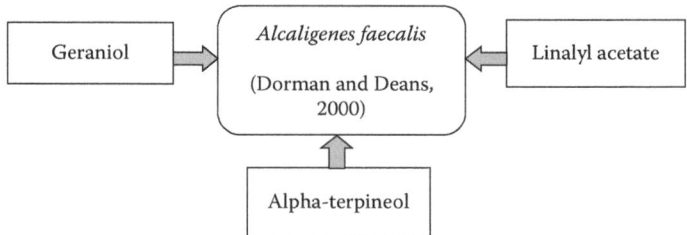

FIGURE 5.2 Natural products interacting with microorganism *Alcaligenes faecalis*.

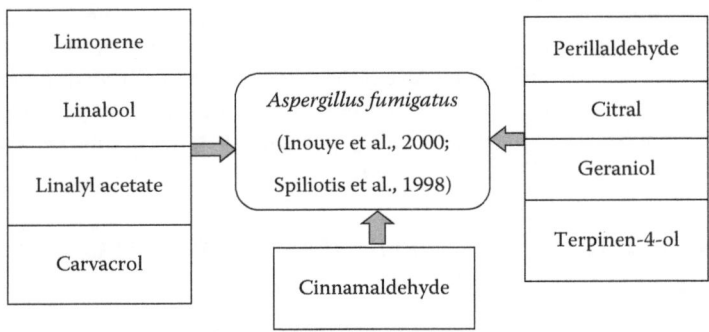

FIGURE 5.3 Natural products interacting with microorganism *Aspergillus fumigatus*.

This organism is a motile, rod shaped, and commonly found in the environment. It is a nonpathogenic and may appear as an opportunistic infection in the urinary tract. These natural products inhibit the population of the *A. faecalis*.

Figure 5.3 introduces nine different natural products that inhibit the microorganism *Aspergillus fumigatus* growth acting on the reproductive cycle genes (Spiliotis et al., 1998; Inouye et al., 2000). The natural products include limonene, linalool, linalyl acetate, carvacrol, cinnamaldehyde, terpinen 4-ol, geraniol, citral, and perillaldehyde. *A. fumigatus* is a saprophyte fungus. It is ubiquitous air-borne fungi. In immunosuppressive patients, the infection of *A. fumigatus* leads to aspergillosis that is fatal and invasive (Latgé, 1999). The mechanism of inhibitory action of these essential oils on apical growth of *A. fumigatus* is through vapor contact. Thus, these compounds in the volatile nature present in the air can suppress the mitotic genes of this fungus apical growth.

Figure 5.4 indicates that four natural products, phenloic glycosides, tannins, linalool, and carvacrol, interact with the genome of of *Aspergillus niger* (Spiliotis et al., 1998; Elgayyar et al., 2001; Hsieh, Mau, and Huang, 2001). This fungus causes a disease called *black mold* on certain fruits and vegetables such as grapes, onions, and peanuts. This is a common contaminant of food. The natural products found to inhibit the growth of this fungus and provide a protection against the infection. Also, this mold rarely reported to cause pneumonia.

Figure 5.5 showed the inhibition of *Bacillus cereus* growth by four natural products, which includes thymoquinone, gamma terpinine, sabinine, and phenolic glycosides

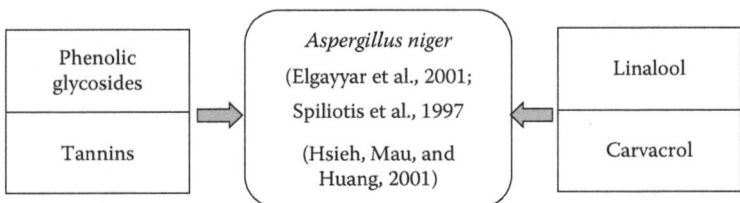

FIGURE 5.4 Natural products interacting with microorganism *Aspergillus niger.*

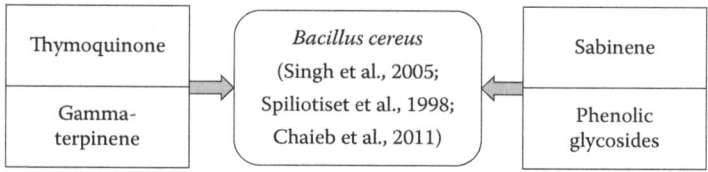

FIGURE 5.5   Natural products interacting with microorganism *Bacillus cereus*.

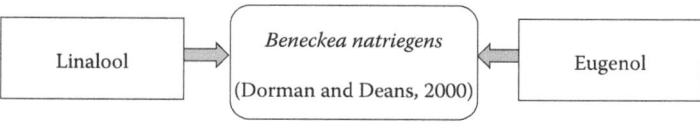

FIGURE 5.6   Natural products interacting with microorganism *Beneckea natriegens*.

(Spiliotis et al., 1998; Singh et al., 2005; Chaieb et al., 2011). This organism produces toxin and grows in cooked food kept a long time at room temperature (http://www. foodsafety.gov). Food including rice, leftovers, sauce, and soups that are kept more than 2 h at room temperature are good sources of this bacterial contamination. Figure 5.6 discusses an interesting organism, *Vibrio natriegens* (previously called as *Beneckea natriegens* and *Pseudomonas natriegens*). This organism has less doubling time when compare to *Escherichia coli* and thus used for many research and experimental settings. This organism is present in salt marsh and has 4788 coding sequences, 14 rRNA coding genes, and 71 tRNA encoding genes. There are two natural products, linalool and eugenol, that inhibit this organism (Dorman and Deans, 2000).

Figure 5.7 documents the effects of natural products on *Bacillus pumilus* organism. This bacterium can survive in prolonged space exposure to ultraviolet

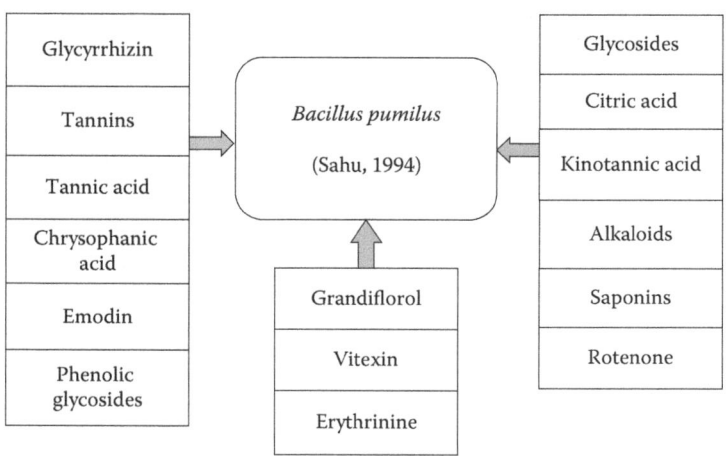

FIGURE 5.7   Natural products interacting with microorganism *Bacillus pumilus*.

radiations due to the accumulation of excess superoxide dismutase (SOD) in the spores to combat the free radicals that arise from the radiations during a space flight to international space station (Vaishampayan et al., 2012). Most importantly, these characteristics of stimulation of SOD further protect its genome from any radiation-induced damage. This organism can even survive in immunosuppressed and cancer patients through catheter. Interestingly, there are 15 natural products that interact with this organism and reduce its growth in spite of its evolved super oxide dismutase protection (Sahu and Chakrabarty, 1994). These natural products are derived from dietary resources, fruits, and vegetables that include glycyrrhizin, tannins, tannic acid, chrysophanic acid, emodin, phenolic glycosides, erythrinine, vitexin, grandiflorol, rotenone, saponins, alkaloids, kinotannic acid, and citric acid.

Figure 5.8 shows 31 natural products interacting with *Bacillus subtilis* bacterium genome and inhibiting its growth. *B. subtilis* lives in soil and in this soil environment for a long time because of the formation of endospores (Eilert, 1981;

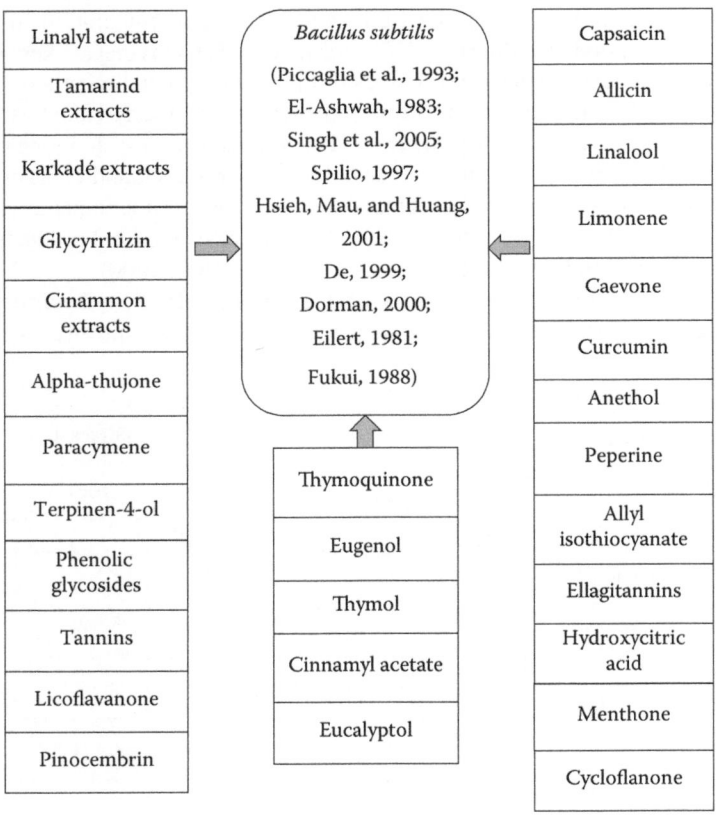

FIGURE 5.8 Natural products interacting with microorganism *Bacillus subtilis.*

El-Ashwah, 1983; Fukui, 1988; Piccaglia et al., 1993; Spiliotis et al., 1998; De, 1999; Dorman, 2000; Hsieh, Mau, and Huang, 2001; Singh et al., 2005). These natural products include linalyl acetate, tamarind extracts, karkadé extracts, glycyrrhizin, cinammon extracts, alpha-thujone, paracymene, terpinen-4-ol, phenolic glycosides, tannins, licoflavanone, pinocembrin, thymoquinone, eugenol, thymol, cinnamyl acetate, eucalyptol, cyclobalanone, menthone, hydroxycitric acid, ellagitannins, allyl isothiocyanate, peperine, anethol, curcumin, caevone, lemonene, linalool, allicin, and capsaicin.

Figure 5.9 shows that the growth of nonsporing bactrerium *Brevibacterium linens* is inhibited by natural products such as paracymene, eucalyptol, alpha-thujone, limonene, estragole, and linalool (Piccaglia et al., 1993; Dorman and Deans, 2000). This organism can grow even at higher salt concentrated foods such as cheese or curd. Generally, this organism is considered a nonpathogenic. Recently, it has been reported that this organism became an opportunistic by CDC group (Gruner et al., 1994). Figure 5.10 shows that five natural products inhibit the growth of *Brochothrix thermosphacta*. This bacterium can grow anaerobically in the higher concentrations of lactate with pH less than 5.5. These natural products include tau-terpinen, beta-thujone, gamma-terpinene, thujone, and borneol. They can grow even in

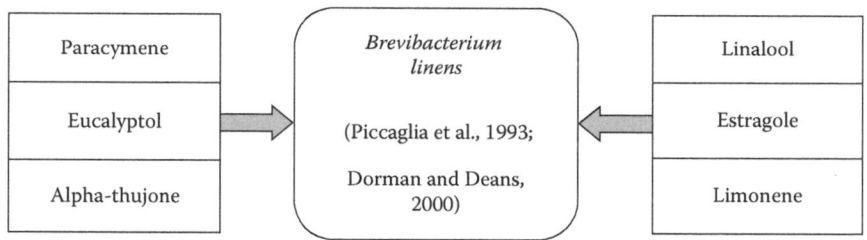

FIGURE 5.9 Natural products interacting with microorganism *Brevibacterium linens*.

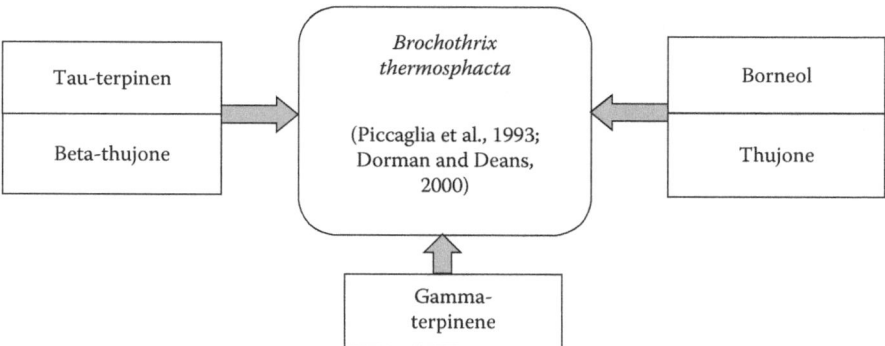

FIGURE 5.10 Natural products interacting with microorganism *Brochothrix thermosphacta*.

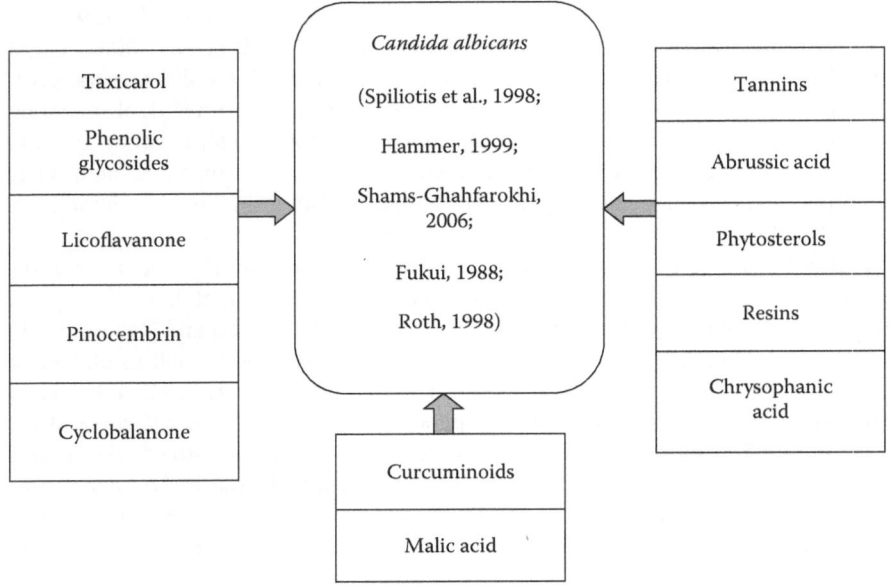

FIGURE 5.11   Natural products interacting with microorganism *Candida albicans*.

vacuum-packaged meat where the lactate is abundant for their growth (Piccaglia et al., 1993; Dorman and Deans, 2000).

Figure 5.11 indicates an interesting organism, *Candida albicans*, that infects gut and induces colonization, which could be a risk factor for triggering food allergy in susceptible individuals (Sugita et al., 2012). There are 12 natural products that could interact and reduce the growth (Fukui, 1988; Spiliotis et al., 1998; Roth, 1998; Hammer, 1999; Shams-Ghahfarokhi et al., 2006). These natural products include taxicarol, phenolic glycosides, licoflavanone, pinocembrin, cyclobalanone, curcuminoids, malic acid, chrysophanic acid, resins, phytosterols, abrussic acid, and tannins. Figure 5.12 demonstrates that there are 30 different natural products that interact with different strains of *E. coli* (DAS, 1957; Eilert, 1981; El-Ashwah, 1983; Kubo, 1991; Spiliotis et al., 1998; Bari, 1999; De, 1999; Hammer, 1999; Dorman and Deans, 2000; Inouye et al., 2000; Elgayyar et al., 2001; Hsieh, Mau, and Huang, 2001; Singh et al., 2005). This is very important in that natural products come from dietary origin and the digestive tract colonization of *E. coli* organisms. These natural products include pterygospermin, calcinated calcium, gamma-terpinene, karkadé extracts, glycyrrhizin, benzoic acid, carvacrol, beta-pinene, capsaicin, limonene, curcumin, punicalagins, hydroxycitric acid, anethol, allyl isothiocyanate, beta-caryophyllene, cinnamaldehyde, geraniol, thymol, neral, peperine, cholind, allicin, eugenol, cinnamyl alcohol, *Rosmarinus officinalis*, phenolic lycosides, *Cinnamomum zeylanicum*, safrole, and thymoquinone.

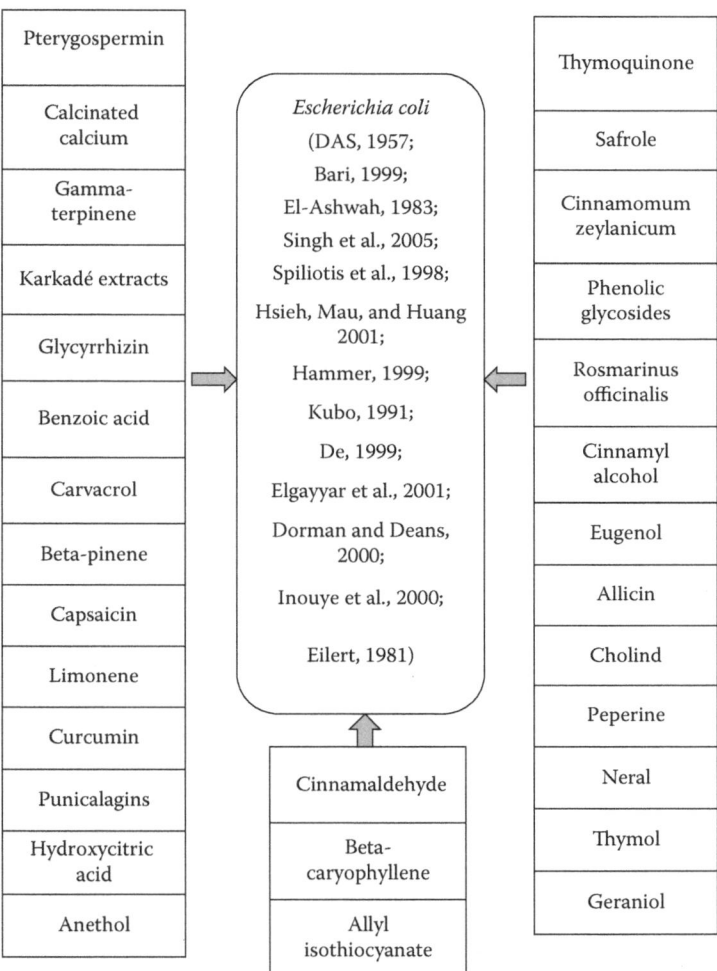

FIGURE 5.12   Natural products interacting with microorganism *Escherichia coli*.

5.3   SINGLE NATURAL PRODUCT INTERACTING WITH ONE OR MORE MICROORGANISMS OR OTHER SPECIES

There are 139 natural products that have been screened using the available literature. They are tabulated for the convenience of discussion as Tables 5.1 through 5.139. The tables are self-explanatory. The main theme of these analyses provides a lot of information about the mechanism of natural product interactions with microbial genomes and a clue when these micobes are present as a community in the micro-biota. We now know which natural product is efficient to inhibit the growth of specific microbes in the microbiota biofilms. In contrast, certain natural products

TABLE 5.1
Interactions of Abrin with Microorganisms

Reference	Plant Name	Phytochemical	Organism
Solis (1969)	*Abrus precatorius*	Abrin	*Sarcina lutea*

TABLE 5.2
Interactions of Abrussic Acid with Microorganisms

Reference	Plant Name	Phytochemical	Organism
Solis (1969)	*Abrus precatorius*	Abrussic acid	*Candida albicans*

TABLE 5.3
Interactions of Ajoene with Microorganisms

Reference	Plant Name	Phytochemical	Organism
Nagourney (1998)	*Allium sativum*	Ajoene	*Salmonella typhimurium*

TABLE 5.4
Interactions of Albuminoids with Microorganisms

Reference	Plant Name	Phytochemical	Organism
Solis (1969)	*Pachyrhizus ersous*	Albuminoids	*Escherichia coli*
Solis (1969)	*Parkia javanica*	Albuminoids	*Escherichia coli*

TABLE 5.5
Interactions of Alkaloids with Microorganisms

Reference	Plant Name	Phytochemical	Organism
Solis (1969)	*Pithecellobium dulce*	Alkaloids	*Escherichia coli*
Solis (1969)	*Samanea saman*	Alkaloids	*Bacillus pumilus*

TABLE 5.6
Interactions of Allicin with Microorganisms

Reference	Plant Name	Phytochemical	Organism
Ody (1993)	*Allium sativum*	Allicin	Methicillin-resistant *Staphylococcus aureus*
Ody (1993)	*Allium sativum*	Allicin	Methicillin-resistant *Staphylococcus aureus*
Nagourney (1998)	*Allium sativum*	Allicin	*Salmonella typhimurium*
De, De, and Banerjiee (1999)	*Allium sativum*	Allicin	*Bacillus subtilis*
De, De, and Banerjiee (1999)	*Allium sativum*	Allicin	*Escherichia coli*
De, De, and Banerjiee (1999)	*Allium sativum*	Allicin	*Saccharomyces cerevisiae*

TABLE 5.7
Interactions of Allyl Isothiocyanate with Microorganisms

Reference	Plant Name	Phytochemical	Organism
De, De, and Banerjiee (1999)	*Armoracia rusticana*	Allyl isothiocyanate	*Bacillus subtilis*
De, De, and Banerjiee (1999)	*Armoracia rusticana*	Allyl isothiocyanate	*Escherichia coli*
De, De, and Banerjiee (1999)	*Armoracia rusticana*	Allyl isothiocyanate	*Saccharomyces cerevisiae*

TABLE 5.8
Interactions of Alpha-Momorcharin with Microorganisms

Reference	Plant Name	Phytochemical	Organism
Zheng, Ben, and Jin (1999)	*Momordica charantia*	Alpha-momorcharin	Human immunodeficiency virus-1

TABLE 5.9
Interactions of Alpha-Basabolol with Microorganisms

Reference	Plant Name	Phytochemical	Organism
Jafri et al. (2001)	*Amomum subulatum*	Alpha-basabolol	*Helicobacter pylori*

TABLE 5.10
Interactions of Alpha-Humulene with Microorganisms

Reference	Plant Name	Phytochemical	Organism
Dorman and Deans (2000)	*Syzygium aromaticum*	Alpha-humulene	*Leuconostoc cremoris*
Dorman and Deans (2000)	*Thymus vulgaris*	Alpha-humulene	*Leuconostoc cremoris*

TABLE 5.11
Interactions of Alpha-Momorcharin with Microorganisms

Reference	Plant Name	Phytochemical	Organism
Mishra and Dubey (1990)	*Momordica charantia*	Alpha-momorcharin	Human immunodeficiency virus

TABLE 5.12
Interactions of Alpha-Phellandrene with Microorganisms

Reference	Plant Name	Phytochemical	Organism
Ruberto et al. (2000)	*Foeniculum vulgare*	Alpha-phellandrene	*Staphylococcus aureus*
Dorman and Deans (2000)	*Syzygium aromaticum*	Alpha-phellandrene	*Lactobacillus plantarum*
Dorman and Deans (2000)	*Thymus vulgaris*	Alpha-phellandrene	*Lactobacillus plantarum*

TABLE 5.13
Interactions of Alpha-Pinene with Microorganisms

Reference	Plant Name	Phytochemical	Organism
Ruberto et al. (2000)	*Foeniculum vulgare*	Alpha-pinene	*Leuconostoc cremoris*
Dorman and Deans (2000)	*Piper nigrum*	Alpha-pinene	*Clostridium sporogenes*
Dorman and Deans (2000)	*Syzygium aromaticum*	Alpha-pinene	*Enterobacter aerogenes*
Dorman and Deans (2000)	*Thymus vulgaris*	Alpha-pinene	*Enterobacter aerogenes*
Jafri et al. (2001)	*Amomum subulatum*	Alpha-pinene	*Helicobacter pylori*
Inouye, Takizawa, and Yamaguchi (2001)	*Rosmarinus officinalis*	Alpha-pinene	*Haemophilus influenzae*
Inouye, Takizawa, and Yamaguchi (2001)	*Rosmarinus officinalis*	Alpha-pinene	*Streptococcus pyogenes*
Inouye, Takizawa, and Yamaguchi (2001)	*Rosmarinus officinalis*	Alpha-pinene	*Streptococcus pneumoniae*
Inouye, Takizawa, and Yamaguchi (2001)	*Rosmarinus officinalis*	Alpha-pinene	*Staphylococcus aureus*
Inouye, Takizawa, and Yamaguchi (2001)	*Rosmarinus officinalis*	Alpha-pinene	*Escherichia coli*

TABLE 5.14
Interactions of Alpha-Terpinene with Microorganisms

Reference	Plant Name	Phytochemical	Organism
Kubo, Himejima, and Muroi (1981)	*Elettaria cardamomum*	Alpha-terpinene	*Brevibacterium ammoniagenes*
Dorman and Deans (2000)	*Myristica fragrans*	Alpha-terpinene	*Micrococcus luteus*

TABLE 5.15
Interactions of Alpha-Terpineol with Microorganisms

Reference	Plant Name	Phytochemical	Organism
Dorman and Deans (2000)	*Myristica fragrans*	Alpha-terpineol	*Alcaligenes faecalis*
Inouye, Takizawa, and Yamaguchi (2001)	*Eucalyptus radiata*	Alpha-terpineol	*Haemophilus influenzae*
Inouye, Takizawa, and Yamaguchi (2001)	*Eucalyptus radiata*	Alpha-terpineol	*Streptococcus pyogenes*
Inouye, Takizawa, and Yamaguchi (2001)	*Eucalyptus radiata*	Alpha-terpineol	*Streptococcus pneumoniae*
Inouye, Takizawa, and Yamaguchi (2001)	*Eucalyptus radiata*	Alpha-terpineol	*Staphylococcus aureus*
Inouye, Takizawa, and Yamaguchi (2001)	*Eucalyptus radiata*	Alpha-terpineol	*Escherichia coli*

TABLE 5.16
Interactions of Alpha-Terpinyl Acetate with Microorganisms

Reference	Plant Name	Phytochemical	Organism
Kubo, Himejima, and Muroi (1981)	*Elettaria cardamomum*	Alpha-terpinyl acetate	*Staphylococcus aureus*

TABLE 5.17
Interactions of Alpha-Thujone with Microorganisms

Reference	Plant Name	Phytochemical	Organism
Piccaglia et al. (1993)	*Salvia officinalis*	Alpha-thujone	*Brevibacterium linens*
Singh et al. (2005)	*Nigella sativa*	Alpha-thujone	*Bacillus subtilis*

TABLE 5.18
Interactions of Anethol with Microorganisms

Reference	Plant Name	Phytochemical	Organism
De, De, and Banerjiee (1999)	*Pimpinella anisum*	Anethol	*Bacillus subtilis*
Elgayyar et al. (2001)	*Pimpinella anisum*	Anethol	*Lactobacillus plantarum*
Elgayyar et al. (2001)	*Pimpinella anisum*	Anethol	*Staphylococcus aureus*
Elgayyar et al. (2001)	*Pimpinella anisum*	Anethol	*Escherichia coli*
Elgayyar et al. (2001)	*Pimpinella anisum*	Anethol	*Salmonella typhimurium*

TABLE 5.19
Interactions of Anise Oil with Microorganisms

Reference	Plant Name	Phytochemical	Organism
Lachowicz et al. (1998)	*Ocimum basilicum*	Anise oil	*Lactobacillus curvatus*
Lachowicz et al. (1998)	*Ocimum basilicum*	Anise oil	*Saccharomyces cerevisiae*

TABLE 5.20
Interactions of Aromadendrene with Microorganisms

Reference	Plant Name	Phytochemical	Organism
Dorman and Deans (2000)	*Piper nigrum*	Aromadendrene	*Erwinia carotovora*
Dorman and Deans (2000)	*Syzygium aromaticum*	Aromadendrene	*Erwinia carotovora*
Dorman and Deans (2000)	*Origanum vulgare*	Aromadendrene	*Erwinia carotovora*
Dorman and Deans (2000)	*Thymus vulgaris*	Aromadendrene	*Erwinia carotovora*

TABLE 5.21
Interactions of Benzoic Acid with Microorganisms

Reference	Plant Name	Phytochemical	Organism
Solis (1969)	*Caesalpinia pulcherrima*	Benzoic acid	*Escherichia coli*

TABLE 5.22
Interactions of Benzylisothiocyanate with Microorganisms

Reference	Plant Name	Phytochemical	Organism
Das, Kurup, and Narasimha Rao (1957)	*Moringa pterygosperma*	Benzylisothiocyanate	*Micrococcus pyogenes* var. *aureus*

TABLE 5.23
Interactions of Beta-Bisabolol with Microorganisms

Reference	Plant Name	Phytochemical	Organism
Dorman and Deans (2000)	*Piper nigrum*	Beta-bisabolol	*Pseudomonas aeruginosa*
Dorman and Deans (2000)	*Syzygium aromaticum*	Beta-bisabolol	*Clostridium sporogenes*
Dorman and Deans (2000)	*Thymus vulgaris*	Beta-bisabolol	*Clostridium sporogenes*

TABLE 5.24
Interactions of Beta-Caryophyllene with Microorganisms

Reference	Plant Name	Phytochemical	Organism
De, De, and Banerjiee (1999)	*Cinnamomum tamala*	Beta-caryophyllene	*Escherichia coli*
Dorman and Deans (2000)	*Piper nigrum*	Beta-caryophyllene	*Escherichia coli*
Dorman and Deans (2000)	*Syzygium aromaticum*	Beta-caryophyllene	*Escherichia coli*
Dorman and Deans (2000)	*Pelargonium graveolens*	Beta-caryophyllene	*Escherichia coli*
Dorman and Deans (2000)	*Origanum vulgare*	Beta-caryophyllene	*Escherichia coli*
Dorman and Deans (2000)	*Thymus vulgaris*	Beta-caryophyllene	*Escherichia coli*

TABLE 5.25
Interactions of Beta-Momorcharin with Microorganisms

Reference	Plant Name	Phytochemical	Organism
Mishra and Dubey (1990)	*Momordica charantia*	Beta-momorcharin	Human immunodeficiency virus

TABLE 5.26
Interactions of Beta-Ocimene with Microorganisms

Reference	Plant Name	Phytochemical	Organism
Dorman and Deans (2000)	*Piper nigrum*	Beta-ocimene	*Enterococcus faecalis*
Dorman and Deans (2000)	*Syzygium aromaticum*	Beta-ocimene	*Enterococcus faecalis*
Dorman and Deans (2000)	*Pelargonium graveolens*	Beta-ocimene	*Enterococcus faecalis*
Dorman and Deans (2000)	*Origanum vulgare*	Beta-ocimene	*Enterococcus faecalis*
Dorman and Deans (2000)	*Thymus vulgaris*	Beta-ocimene	*Enterococcus faecalis*

TABLE 5.27
Interactions of Beta-Phellandrene with Microorganisms

Reference	Plant Name	Phytochemical	Organism
Singh et al. (2005)	*Myristica fragrans*	Beta-phellandrene	*Salmonella typhi*

TABLE 5.28
Interactions of Beta-Pinene with Microorganisms

Reference	Plant Name	Phytochemical	Organism
Singh et al. (2005)	*Cuminum cyminum*	Beta-pinene	*Escherichia coli*
De, De, and Banerjiee (1999)	*Cuminum cyminum*	Beta-pinene	*Escherichia coli*
De, De, and Banerjiee (1999)	*Trachyspermum ammi*	Beta-pinene	*Escherichia coli*
Dorman and Deans (2000)	*Myristica fragrans*	Beta-pinene	*Klebsiella pneumoniae*

TABLE 5.29
Interactions of Beta-Thujone with Microorganisms

Reference	Plant Name	Phytochemical	Organism
Piccaglia et al. (1993)	*Salvia officinalis*	Beta-thujone	*Brochothrix thermosphacta*

TABLE 5.30
Interactions of Beta-Trichosanthin with Microorganisms

Reference	Plant Name	Phytochemical	Organism
Mishra, and Dubey (1990)	*Trichosanthes cucumerina*	Beta-trichosanthin	Human immunodeficiency virus

TABLE 5.31
Interactions of Bicalin with Microorganisms

Reference	Plant Name	Phytochemical	Organism
Yamashiki et al. (1999)	*Scutellaria*	Bicalin	Flaviviridae

TABLE 5.32
Interactions of Borneol with Microorganisms

Reference	Plant Name	Phytochemical	Organism
Dorman and Deans (2000)	*Piper nigrum*	Borneol	*Brochothrix thermosphacta*
Dorman and Deans (2000)	*Syzygium aromaticum*	Borneol	*Brochothrix thermosphacta*
Dorman and Deans (2000)	*Pelargonium graveolens*	Borneol	*Brochothrix thermosphacta*
Dorman and Deans (2000)	*Myristica fragrans*	Borneol	*Moraxella* sp.
Dorman and Deans (2000)	*Origanum vulgare*	Borneol	*Brochothrix thermosphacta*
Dorman and Deans (2000)	*Thymus vulgaris*	Borneol	*Brochothrix thermosphacta*

TABLE 5.33
Interactions of Bornyl Acetate with Microorganisms

Reference	Plant Name	Phytochemical	Organism
Dorman and Deans (2000)	*Piper nigrum*	Bornyl acetate	*Leuconostoc cremoris*
Dorman and Deans (2000)	*Syzygium aromaticum*	Bornyl acetate	*Klebsiella pneumoniae*
Dorman and Deans (2000)	*Pelargonium graveolens*	Bornyl acetate	*Klebsiella pneumoniae*
Dorman and Deans (2000)	*Myristica fragrans*	Bornyl acetate	*Serratia marcescens*
Dorman and Deans (2000)	*Origanum vulgare*	Bornyl acetate	*Klebsiella pneumoniae*
Dorman and Deans (2000)	*Thymus vulgaris*	Bornyl acetate	*Klebsiella pneumoniae*

TABLE 5.34
Interactions of Brazilin with Microorganisms

Reference	Plant Name	Phytochemical	Organism
Solis (1969)	*Caesalpinia sappan*	Brazilin	*Escherichia coli*

TABLE 5.35
Interactions of Caevone with Microorganisms

Reference	Plant Name	Phytochemical	Organism
De, De, and Banerjiee (1999)	*Carum carvi*	Caevone	*Bacillus subtilis*

TABLE 5.36
Interactions of Calcinated Calcium with Microorganisms

Reference	Plant Name	Phytochemical	Organism
Barl et al. (1999)	*Raphanus sativus*	Calcinated calcium	*Escherichia coli*

TABLE 5.37
Interactions of Camphor with Microorganisms

Reference	Plant Name	Phytochemical	Organism
Piccaglia et al. (1993)	*Lavandula angustifolia*	Camphor	*Staphylococcus aureus*
Piccaglia et al. (1993)	*Salvia officinalis*	Camphor	*Clostridium sporogenes*
Inouye, Takizawa, and Yamaguchi (2001)	*Lavandula latifolia*	Camphor	*Haemophilus influenzae*
Inouye, Takizawa, and Yamaguchi (2001)	*Lavandula latifolia*	Camphor	*Streptococcus pyogenes*
Inouye, Takizawa, and Yamaguchi (2001)	*Lavandula latifolia*	Camphor	*Streptococcus pneumoniae*
Inouye, Takizawa, and Yamaguchi (2001)	*Lavandula latifolia*	Camphor	*Staphylococcus aureus*
Inouye, Takizawa, and Yamaguchi (2001)	*Lavandula latifolia*	Camphor	*Escherichia coli*
Inouye, Takizawa, and Yamaguchi (2001)	*Rosmarinus officinalis*	Camphor	*Haemophilus influenzae*
Inouye, Takizawa, and Yamaguchi (2001)	*Rosmarinus officinalis*	Camphor	*Streptococcus pyogenes*
Inouye, Takizawa, and Yamaguchi (2001)	*Rosmarinus officinalis*	Camphor	*Streptococcus pneumoniae*
Inouye, Takizawa, and Yamaguchi (2001)	*Rosmarinus officinalis*	Camphor	*Staphylococcus aureus*
Inouye, Takizawa, and Yamaguchi (2001)	*Rosmarinus officinalis*	Camphor	*Escherichia coli*

TABLE 5.38
Interactions of Capsaicin with Microorganisms

Reference	Plant Name	Phytochemical	Organism
De, De, and Banerjiee (1999)	*Capsicum annum*	Capsaicin	*Bacillus subtilis*
De, De, and Banerjiee (1999)	*Capsicum annum*	Capsaicin	*Escherichia coli*
De, De, and Banerjiee (1999)	*Capsicum annum*	Capsaicin	*Saccharomyces cerevisiae*

TABLE 5.39
Interactions of Carene with Microorganisms

Reference	Plant Name	Phytochemical	Organism
Dorman and Deans (2000)	*Piper nigrum*	Carene	*Yersinia enterocolitica*
Dorman and Deans (2000)	*Syzygium aromaticum*	Carene	*Yersinia enterocolitica*
Dorman and Deans (2000)	*Myristica fragrans*	Carene	*Citrobacter freundii*
Dorman and Deans (2000)	*Origanum vulgare*	Carene	*Yersinia enterocolitica*
Dorman and Deans (2000)	*Thymus vulgaris*	Carene	*Yersinia enterocolitica*

TABLE 5.40
Interactions of Carvacrol with Microorganisms

Reference	Plant Name	Phytochemical	Organism
Piccaglia et al. (1993)	*Satureja montana*	Carvacrol	*Lactobacillus plantarum*
Inouye et al. (2000)	*Thymus serpyllum*	Carvacrol	*Aspergillus fumigatus*
Singh et al. (2005)	*Nigella sativa*	Carvacrol	*Escherichia coli*
Elgayyar et al. (2001)	*Origanum vulgare*	Carvacrol	*Lactobacillus plantarum*
Elgayyar et al. (2001)	*Origanum vulgare*	Carvacrol	*Rhodotorula*
Elgayyar et al. (2001)	*Origanum vulgare*	Carvacrol	*Aspergillus niger*
Dorman and Deans (2000)	*Piper nigrum*	Carvacrol	*Staphylococcus aureus*
Dorman and Deans (2000)	*Syzygium aromaticum*	Carvacrol	*Staphylococcus aureus*
Dorman and Deans (2000)	*Pelargonium graveolens*	Carvacrol	*Staphylococcus aureus*
Dorman and Deans (2000)	*Myristica fragrans*	Carvacrol	*Aeromonas hydrophila*
Dorman and Deans (2000)	*Origanum vulgare*	Carvacrol	*Staphylococcus aureus*
Dorman and Deans (2000)	*Thymus vulgaris*	Carvacrol	*Staphylococcus aureus*
Inouye, Takizawa, and Yamaguchi (2001)	*Thymus serpyllum*	Carvacrol	*Haemophilus influenzae*
Inouye, Takizawa, and Yamaguchi (2001)	*Thymus serpyllum*	Carvacrol	*Streptococcus pyogenes*
Inouye, Takizawa, and Yamaguchi (2001)	*Thymus serpyllum*	Carvacrol	*Streptococcus pneumoniae*
Inouye, Takizawa, and Yamaguchi (2001)	*Thymus serpyllum*	Carvacrol	*Staphylococcus aureus*
Inouye, Takizawa, and Yamaguchi (2001)	*Thymus serpyllum*	Carvacrol	*Escherichia coli*

TABLE 5.41
Interactions of Carvacrol Methyl Ether with Microorganisms

Reference	Plant Name	Phytochemical	Organism
Dorman and Deans (2000)	*Myristica fragrans*	Carvacrol methyl ether	*Staphylococcus aureus*

TABLE 5.42
Interactions of Chinese Chive with Microorganisms

Reference	Plant Name	Phytochemical	Organism
Hsieh, Mau, and Huang (2001)	*Allium tuberosum*	Chinese chive	*Listeria monocytogenes*
Hsieh, Mau, and Huang (2001)	*Allium tuberosum*	Chinese chive	*Vibrio parahaemolyticus*
Hsieh, Mau, and Huang (2001)	*Allium tuberosum*	Chinese chive	*Aspergillus flavus*

TABLE 5.43
Interactions of Chlorogenic Acid with Microorganisms

Reference	Plant Name	Phytochemical	Organism
Mansour and Khalil (2000)	*Solanum tuberosum*	Chlorogenic acid	Antioxidant against beef patties

TABLE 5.44
Interactions of Cholind with Microorganisms

Reference	Plant Name	Phytochemical	Organism
De, De, and Banerjiee (1999)	*Pimpinella anisum*	Cholind	*Escherichia coli*

TABLE 5.45
Interactions of Chrysophanic Acid with Microorganisms

Reference	Plant Name	Phytochemical	Organism
Solis (1969)	*Cassia alata*	Chrysophanic acid	*Bacillus pumilus*
Solis (1969)	*Cassia tora*	Chrysophanic acid	*Candida albicans*

TABLE 5.46
Interactions of Cinammon Extracts with Microorganisms

Reference	Plant Name	Phytochemical	Organism
Alian, El-Ashwah, and Eid (1983)	*Cinnamon zeylanicum*	Cinammon extracts	*Escherichia coli*
Alian, El-Ashwah, and Eid (1983)	*Cinnamon zeylanicum*	Cinammon extracts	*Pseudomonas pyocyaneus*
Alian, El-Ashwah, and Eid (1983)	*Cinnamon zeylanicum*	Cinammon extracts	*Salmonella typhi*
Alian, El-Ashwah, and Eid (1983)	*Cinnamon zeylanicum*	Cinammon extracts	*Klebsiella pneumoniae*
Alian, El-Ashwah, and Eid (1983)	*Cinnamon zeylanicum*	Cinammon extracts	*Bacillus subtilis*
De, De, and Banerjiee (1999)	*Cinnamon zeylanicum*	Cinammon extracts	*Saccharomyces cerevisiae*

TABLE 5.47
Interactions of Cinnamaldehyde with Microorganisms

Reference	Plant Name	Phytochemical	Organism
Inouye et al. (2000)	*Cinnamomum verum*	Cinnamaldehyde	*Aspergillus fumigatus*
Hsieh, Mau, and Huang (2001)	*Cinnamomum cassia*	Cinnamaldehyde	*Escherichia coli*
Hsieh, Mau, and Huang (2001)	*Cinnamomum cassia*	Cinnamaldehyde	*Salmonella typhimurium*
Hsieh, Mau, and Huang (2001)	*Cinnamomum cassia*	Cinnamaldehyde	*Kloeckera apiculata*
Hsieh, Mau, and Huang (2001)	*Cinnamomum cassia*	Cinnamaldehyde	*Aureobasidium pullulans*
Inouye, Takizawa, and Yamaguchi (2001)	*Cinnamomum zeylanicum*	Cinnamaldehyde	*Haemophilus influenzae*
Inouye, Takizawa, and Yamaguchi (2001)	*Cinnamomum zeylanicum*	Cinnamaldehyde	*Streptococcus pyogenes*
Inouye, Takizawa, and Yamaguchi (2001)	*Cinnamomum zeylanicum*	Cinnamaldehyde	*Streptococcus pneumoniae*
Inouye, Takizawa, and Yamaguchi (2001)	*Cinnamomum zeylanicum*	Cinnamaldehyde	*Staphylococcus aureus*
Inouye, Takizawa, and Yamaguchi (2001)	*Cinnamomum zeylanicum*	Cinnamaldehyde	*Escherichia coli*

TABLE 5.48
Interactions of Cinnamyl Acetate with Microorganisms

Reference	Plant Name	Phytochemical	Organism
De, De, and Banerjiee (1999)	*Cinnamon zeylanicum*	Cinnamyl acetate	*Bacillus subtilis*

TABLE 5.49
Interactions of Cinnamyl Alcohol with Microorganisms

Reference	Plant Name	Phytochemical	Organism
De, De, and Banerjiee (1999)	*Cinnamon zeylanicum*	Cinnamyl alcohol	*Escherichia coli*

TABLE 5.50
Citral Interactions on Microorganisms

Reference	Plant Name	Phytochemical	Organism
Inouye et al. (2000)	*Cymbopogon citratus*	Citral	*Aspergillus fumigatus*
Dorman and Deans (2000)	*Piper nigrum*	Citral	*Acinetobacter calcoaceticus*
Dorman and Deans (2000)	*Syzygium aromaticum*	Citral	*Acinetobacter calcoaceticus*
Dorman and Deans (2000)	*Pelargonium graveolens*	Citral	*Acinetobacter calcoaceticus*
Dorman and Deans (2000)	*Origanum vulgare*	Citral	*Acinetobacter calcoaceticus*
Dorman and Deans (2000)	*Thymus vulgaris*	Citral	*Acinetobacter calcoaceticus*

TABLE 5.51
Interactions of Citric Acid with Microorganisms

Reference	Plant Name	Phytochemical	Organism
Solis (1969)	*Cassia fistula*	Citric acid	*Sarcina lutea*
Solis (1969)	*Parkia javanica*	Citric acid	*Bacillus pumilus*

TABLE 5.52
Interactions of Coumarines with Microorganisms

Reference	Plant Name	Phytochemical	Organism
Lans and Brown (1998)	*Petiveria alliacea*	Coumarines	To promote chicken health

TABLE 5.53
Interactions of Cuminaldehyde with Microorganisms

Reference	Plant Name	Phytochemical	Organism
Singh et al. (2005)	*Cuminum cyminum*	Cuminaldehyde	*Staphylococcus aureus*
De, De, and Banerjiee (1999)	*Cuminum cyminum*	Cuminaldehyde	*Saccharomyces cerevisiae*

TABLE 5.54
Interactions of Curcumin with Microorganisms

Reference	Plant Name	Phytochemical	Organism
De, De, and Banerjiee (1999)	*Curcuma longa*	Curcumin	*Bacillus subtilis*
De, De, and Banerjiee (1999)	*Curcuma longa*	Curcumin	*Escherichia coli*
De, De, and Banerjiee (1999)	*Curcuma longa*	Curcumin	*Saccharomyces cerevisiae*

TABLE 5.55
Interactions of Curcuminoids with Microorganisms

Reference	Plant Name	Phytochemical	Organism
Roth, Chandra, and Nair (1998)	*Curcuma longa*	Curcuminoids	*Candida albicans*
Roth, Chandra, and Nair (1998)	*Curcuma longa*	Curcuminoids	*Candida krusei*
Roth, Chandra, and Nair (1998)	*Curcuma longa*	Curcuminoids	*Candida parapsilosis*

TABLE 5.56
Interactions of Cycloflanone with Microorganisms

Reference	Plant Name	Phytochemical	Organism
Fukui, Goto, and Tabata (1988)	*Glycyrrhiza glabra*	Cycloflanone	*Bacillus subtilis*
Fukui, Goto, and Tabata (1988)	*Glycyrrhiza glabra*	Cycloflanone	*Staphylococcus aureus*
Fukui, Goto, and Tabata (1988)	*Glycyrrhiza glabra*	Cycloflanone	*Candida albicans*

TABLE 5.57
Interactions of D-Camphene with Microorganisms

Reference	Plant Name	Phytochemical	Organism
De, De, and Banerjiee (1999)	*Myristica fragrans*	D-Camphene	*Saccharomyces cerevisiae*

TABLE 5.58
Interactions of Derrin with Microorganisms

Reference	Plant Name	Phytochemical	Organism
Solis (1969)	*Derris elliptica*	Derrin	*Sarcina lutea*

TABLE 5.59
Interactions of Dianthin 32 with Microorganisms

Reference	Plant Name	Phytochemical	Organism
Tomasi et al. (1982)	*Dianthus caryophyllus*	Dianthin 32	Poliovirus

TABLE 5.60
Interactions of Ellagitannins with Microorganisms

Reference	Plant Name	Phytochemical	Organism
De, De, and Banerjiee (1999)	*Punica granatum*	Ellagitannins	*Bacillus subtilis*

TABLE 5.61
Interactions of Emodin with Microorganisms

Reference	Plant Name	Phytochemical	Organism
Solis (1969)	*Cassia tora*	Emodin	*Bacillus pumilus*

TABLE 5.62
Interactions of Erythrinine with Microorganisms

Reference	Plant Name	Phytochemical	Organism
Solis (1969)	*Erythrina variegata* var. *orientalis*	Erythrinine	*Bacillus pumilus*

TABLE 5.63
Interactions of Estragole with Microorganisms

Reference	Plant Name	Phytochemical	Organism
Ruberto et al. (2000)	*Foeniculum vulgare*	Estragole	*Brevibacterium linens*
Ruberto et al. (2000)	*Foeniculum vulgare*	Estragole	*Clostridium perfringens*

TABLE 5.64
Interactions of Eucalyptol with Microorganisms

Reference	Plant Name	Phytochemical	Organism
Piccaglia et al. (1993)	*Lavandula angustifolia*	Eucalyptol	*Brevibacterium linens*
Piccaglia et al. (1993)	*Salvia officinalis*	Eucalyptol	*Staphylococcus aureus*
Kubo, Himejima, and Muroi (1981)	*Elettaria sardamomum*	Eucalyptol	*Bacillus subtilis*
Dorman and Deans (2000)	*Piper nigrum*	Eucalyptol	*Flavobacterium suaveolens*
Dorman and Deans (2000)	*Syzygium aromaticum*	Eucalyptol	*Flavobacterium suaveolens*
Dorman and Deans (2000)	*Pelargonium graveolens*	Eucalyptol	*Flavobacterium suaveolens*
Dorman and Deans (2000)	*Myristica fragrans*	Eucalyptol	*Salmonella pullorum*
Dorman and Deans (2000)	*Origanum vulgare*	Eucalyptol	*Flavobacterium suaveolens*
Dorman and Deans (2000)	*Thymus vulgaris*	Eucalyptol	*Flavobacterium suaveolens*
Jafri et al. (2001)	*Amomum subulatum*	Eucalyptol	*Helicobacter pylori*
Inouye, Takizawa, and Yamaguchi (2001)	*Lavandula latifolia*	Eucalyptol	*Haemophilus influenzae*
Inouye, Takizawa, and Yamaguchi (2001)	*Lavandula latifolia*	Eucalyptol	*Streptococcus pyogenes*
Inouye, Takizawa, and Yamaguchi (2001)	*Lavandula latifolia*	Eucalyptol	*Streptococcus pneumoniae*

(Continued)

TABLE 5.64 (*Continued*)
Interactions of Eucalyptol with Microorganisms

Reference	Plant Name	Phytochemical	Organism
Inouye, Takizawa, and Yamaguchi (2001)	*Lavandula latifolia*	Eucalyptol	*Staphylococcus aureus*
Inouye, Takizawa, and Yamaguchi (2001)	*Lavandula latifolia*	Eucalyptol	*Escherichia coli*
Inouye, Takizawa, and Yamaguchi (2001)	*Rosmarinus officinalis*	Eucalyptol	*Haemophilus influenzae*
Inouye, Takizawa, and Yamaguchi (2001)	*Rosmarinus officinalis*	Eucalyptol	*Streptococcus pyogenes*
Inouye, Takizawa, and Yamaguchi (2001)	*Rosmarinus officinalis*	Eucalyptol	*Streptococcus pneumoniae*
Inouye, Takizawa, and Yamaguchi (2001)	*Rosmarinus officinalis*	Eucalyptol	*Staphylococcus aureus*
Inouye, Takizawa, and Yamaguchi (2001)	*Rosmarinus officinalis*	Eucalyptol	*Escherichia coli*
Inouye, Takizawa, and Yamaguchi (2001)	*Eucalyptus radiata*	Eucalyptol	*Haemophilus influenzae*
Inouye, Takizawa, and Yamaguchi (2001)	*Eucalyptus radiata*	Eucalyptol	*Streptococcus pyogenes*
Inouye, Takizawa, and Yamaguchi (2001)	*Eucalyptus radiata*	Eucalyptol	*Streptococcus pneumoniae*
Inouye, Takizawa, and Yamaguchi (2001)	*Eucalyptus radiata*	Eucalyptol	*Staphylococcus aureus*
Inouye, Takizawa, and Yamaguchi (2001)	*Eucalyptus radiata*	Eucalyptol	*Escherichia coli*

TABLE 5.65
Interactions of Eugenol with Microorganisms

Reference	Plant Name	Phytochemical	Organism
Hsieh, Mau, and Huang (2001)	*Cinnamomum cassia*	Eugenol	*Flavobacterium suaveolens*
Hsieh, Mau, and Huang (2001)	*Cinnamomum cassia*	Eugenol	*Staphylococcus aureus*
Hsieh, Mau, and Huang (2001)	*Cinnamomum cassia*	Eugenol	*Pichia membranaefaciens*
Hsieh, Mau, and Huang (2001)	*Cinnamomum cassia*	Eugenol	*Penicillium italicum*
Kubo, Himejima, and Muroi (1981)	*Elettaria cardamomum*	Eugenol	*Pseudomonas aeruginosa*
Kubo, Himejima, and Muroi (1981)	*Elettaria cardamomum*	Eugenol	*Candida utilis*

(Continued)

TABLE 5.65 (*Continued*)
Interactions of Eugenol with Microorganisms

Reference	Plant Name	Phytochemical	Organism
De, De, and Banerjiee (1999)	*Eugenia caryophyllus*	Eugenol	*Bacillus subtilis*
De, De, and Banerjiee (1999)	*Eugenia caryophyllus*	Eugenol	*Escherichia coli*
De, De, and Banerjiee (1999)	*Eugenia caryophyllus*	Eugenol	*Saccharomyces cerevisiae*
De, De, and Banerjiee (1999)	*Ocimum sanctum*	Eugenol	*Saccharomyces cerevisiae*
De, De, and Banerjiee (1999)	*Cinnamomum tamala*	Eugenol	*Saccharomyces cerevisiae*
Elgayyar et al. (2001)	*Ocimum basilicum*	Eugenol	*Staphylococcus aureus*
Elgayyar et al. (2001)	*Ocimum basilicum*	Eugenol	*Yersinia enterocolitica*
Elgayyar et al. (2001)	*Ocimum basilicum*	Eugenol	*Escherichia coli*
Elgayyar et al. (2001)	*Ocimum basilicum*	Eugenol	*Salmonella typhimurium*
Elgayyar et al. (2001)	*Ocimum basilicum*	Eugenol	*Listeria monocytogenes*
Elgayyar et al. (2001)	*Ocimum basilicum*	Eugenol	*Pseudomonas aeruginosa*
Dorman and Deans (2000)	*Piper nigrum*	Eugenol	*Proteus vulgaris*
Dorman and Deans (2000)	*Syzygium aromaticum*	Eugenol	*Proteus vulgaris*
Dorman and Deans (2000)	*Myristica fragrans*	Eugenol	*Beneckea natriegens*
Dorman and Deans (2000)	*Origanum vulgare*	Eugenol	*Proteus vulgaris*
Dorman and Deans (2000)	*Thymus vulgaris*	Eugenol	*Proteus vulgaris*
Lans and Brown (1998)	*Pimenta racemosa*	Eugenol	To promote chicken health

TABLE 5.66
Interactions of Fenchone as Anti-Oxidant Activity

Reference	Plant Name	Phytochemical	Activity
Guillen and Manzanos (1996)	*Foeniculum vulgare*	Fenchone	Antioxidant activity in vitro

TABLE 5.67
Interactions of Gallic Acid with Microorganisms

Reference	Plant Name	Phytochemical	Organism
Solis (1969)	*Caesalpinia pulcherrima*	Gallic acid	*Sarcina lutea*
Solis (1969)	*Caesalpinia sappan*	Gallic acid	*Sarcina lutea*

TABLE 5.68
Interactions of Gamma-Terpinene with Microorganisms

Reference	Plant Name	Phytochemical	Organism
Piccaglia et al. (1993)	*Satureja montana*	Gamma-terpinene	*Brochothrix thermosphacta*
Ruberto et al. (2000)	*Crithmum maritimum*	Gamma-terpinene	*Clostridium perfringens*
Singh et al. (2005)	*Cuminum cyminum*	Gamma-terpinene	*Bacillus cereus*
Dorman and Deans (2000)	*Piper nigrum*	Gamma-terpinene	*Micrococcus luteus*
Dorman and Deans (2000)	*Syzygium aromaticum*	Gamma-terpinene	*Micrococcus luteus*
Dorman and Deans (2000)	*Pelargonium graveolens*	Gamma-terpinene	*Micrococcus luteus*
Dorman and Deans (2000)	*Origanum vulgare*	Gamma-terpinene	*Micrococcus luteus*
Dorman and Deans (2000)	*Thymus vulgaris*	Gamma-terpinene	*Micrococcus luteus*
Inouye, Takizawa, and Yamaguchi (2001)	*Thymus vulgaris*	Gamma-terpinene	*Haemophilus influenzae*
Inouye, Takizawa, and Yamaguchi (2001)	*Thymus vulgaris*	Gamma-terpinene	*Streptococcus pyogenes*
Inouye, Takizawa, and Yamaguchi (2001)	*Thymus vulgaris*	Gamma-terpinene	*Streptococcus pneumoniae*
Inouye, Takizawa, and Yamaguchi (2001)	*Thymus vulgaris*	Gamma-terpinene	*Staphylococcus aureus*
Inouye, Takizawa, and Yamaguchi (2001)	*Thymus vulgaris*	Gamma-terpinene	*Escherichia coli*
Inouye, Takizawa, and Yamaguchi (2001)	*Melaleuca alternifolia*	Gamma-terpinene	*Haemophilus influenzae*
Inouye, Takizawa, and Yamaguchi (2001)	*Melaleuca alternifolia*	Gamma-terpinene	*Streptococcus pyogenes*
Inouye, Takizawa, and Yamaguchi (2001)	*Melaleuca alternifolia*	Gamma-terpinene	*Streptococcus pneumoniae*
Inouye, Takizawa, and Yamaguchi (2001)	*Melaleuca alternifolia*	Gamma-terpinene	*Staphylococcus aureus*
Inouye, Takizawa, and Yamaguchi (2001)	*Melaleuca alternifolia*	Gamma-terpinene	*Escherichia coli*

TABLE 5.69
Interactions of Gelonin with Microorganisms

Reference	Plant Name	Phytochemical	Organism
Tomasi et al. (1982)	*Gelonium multiflorum*	Gelonin	Herpes simplex virus-1

TABLE 5.70
Interactions of Geraniol with Microorganisms

Reference	Plant Name	Phytochemical	Organism
Inouye et al. (2000)	*Cymbopogon citratus*	Geraniol	*Aspergillus fumigatus*
Kubo, Himejima, and Muroi (1981)	*Elettaria cardamomum*	Geraniol	*Streptococcus mutans*
Dorman and Deans (2000)	*Piper nigrum*	Geraniol	*Alcaligenes faecalis*
Dorman and Deans (2000)	*Syzygium aromaticum*	Geraniol	*Alcaligenes faecalis*
Dorman and Deans (2000)	*Myristica fragrans*	Geraniol	*Escherichia coli*
Dorman and Deans (2000)	*Origanum vulgare*	Geraniol	*Alcaligenes faecalis*
Dorman and Deans (2000)	*Thymus vulgaris*	Geraniol	*Alcaligenes faecalis*
Inouye, Takizawa, and Yamaguchi (2001)	*Cymbopogon*	Geraniol	*Haemophilus influenzae*
Inouye, Takizawa, and Yamaguchi (2001)	*Cymbopogon*	Geraniol	*Streptococcus pyogenes*
Inouye, Takizawa, and Yamaguchi (2001)	*Cymbopogon*	Geraniol	*Streptococcus pneumoniae*
Inouye, Takizawa, and Yamaguchi (2001)	*Cymbopogon*	Geraniol	*Staphylococcus aureus*
Inouye, Takizawa, and Yamaguchi (2001)	*Cymbopogon*	Geraniol	*Escherichia coli*
Inouye, Takizawa, and Yamaguchi (2001)	*Thymus vulgaris* ct. *geraniol*	Geraniol	*Haemophilus influenzae*
Inouye, Takizawa, and Yamaguchi (2001)	*Thymus vulgaris* ct. *geraniol*	Geraniol	*Streptococcus pyogenes*
Inouye, Takizawa, and Yamaguchi (2001)	*Thymus vulgaris* ct. *geraniol*	Geraniol	*Streptococcus pneumoniae*
Inouye, Takizawa, and Yamaguchi (2001)	*Thymus vulgaris* ct. *geraniol*	Geraniol	*Staphylococcus aureus*
Inouye, Takizawa, and Yamaguchi (2001)	*Thymus vulgaris* ct. *geraniol*	Geraniol	*Escherichia coli*

TABLE 5.71
Interactions of Geranyl Acetate with Microorganisms

Reference	Plant Name	Phytochemical	Organism
Dorman and Deans (2000)	*Myristica fragrans*	Geranyl acetate	*Enterococcus faecalis*
Inouye, Takizawa, and Yamaguchi (2001)	*Thymus vulgaris* ct. *geraniol*	Geranyl acetate	*Haemophilus influenzae*
Inouye, Takizawa, and Yamaguchi (2001)	*Thymus vulgaris* ct. *geraniol*	Geranyl acetate	*Streptococcus pyogenes*
Inouye, Takizawa, and Yamaguchi (2001)	*Thymus vulgaris* ct. *geraniol*	Geranyl acetate	*Streptococcus pneumoniae*
Inouye, Takizawa, and Yamaguchi (2001)	*Thymus vulgaris* ct. *geraniol*	Geranyl acetate	*Staphylococcus aureus*
Inouye, Takizawa, and Yamaguchi (2001)	*Thymus vulgaris* ct. *geraniol*	Geranyl acetate	*Escherichia coli*

TABLE 5.72
Interactions of Gingerols as an Antioxidant Activity

Reference	Plant Name	Phytochemical	Activity
Mansour and Khalil (2000)	*Zingiber officinale*	Gingerols	Antioxidant

TABLE 5.73
Interactions of Glycerides with Microorganisms

Reference	Plant Name	Phytochemical	Organism
Solis (1969)	*Pithecellobium dulce*	Gycerides	*Bacillus pumilus*

TABLE 5.74
Interactions of Aromatic Glycosides with Microorganisms

Reference	Plant Name	Phytochemical	Organism
Lans and Brown (1998)	*Momordica charantia*	Glycosides	To promote chicken heath
Eilbert, Wolters, and Nahrstedt (1981)	*Moringa oleifera*	Glycosides	*Bacillus subtilis*
Eilbert, Wolters, and Nahrstedt (1981)	*Moringa oleifera*	Glycosides	*Mycobacterium phlei*
Eilbert, Wolters, and Nahrstedt (1981)	*Moringa oleifera*	Glycosides	*Escherichia coli*
Eilbert, Wolters, and Nahrstedt (1981)	*Moringa oleifera*	Glycosides	*Staphylococcus aureus*
Eilbert, Wolters, and Nahrstedt (1981)	*Moringa oleifera*	Glycosides	*Proteus mirabilis*
Eilbert, Wolters, and Nahrstedt (1981)	*Moringa oleifera*	Glycosides	*Salmonella edinburg*
Eilbert, Wolters, and Nahrstedt (1981)	*Moringa oleifera*	Glycosides	*Serratia marcescens*
Eilbert, Wolters, and Nahrstedt (1981)	*Moringa oleifera*	Glycosides	*Streptococcus faecalis*
Eilbert, Wolters, and Nahrstedt (1981)	*Moringa oleifera*	Glycosides	*Candida pseudotropicalis*
Eilbert, Wolters, and Nahrstedt (1981)	*Moringa stenopetala*	Glycosides	*Bacilluss*
Eilbert, Wolters, and Nahrstedt (1981)	*Moringa stenopetala*	Glycosides	*Mycobacterium phlei*
Solis (1969)	*Mimosa pudica*	Glycosides	*Escherichia coli*
Solis (1969)	*Pachyrhizus erosus*	Glycosides	*Bacillus pumilus*

TABLE 5.75
Interactions of Glycyrrhizin with Microorganisms

Reference	Plant Name	Phytochemical	Organism
Alian, El-Ashwah, and Eid (1983)	*Glycyrrhiza glabra*	Glycyrrhizin	*Escherichia coli*
Alian, El-Ashwah, and Eid (1983)	*Glycyrrhiza glabra*	Glycyrrhizin	*Pseudomonas pyocyaneus*
Alian, El-Ashwah, and Eid (1983)	*Glycyrrhiza glabra*	Glycyrrhizin	*Salmonella typhi*
Alian, El-Ashwah, and Eid (1983)	*Glycyrrhiza glabra*	Glycyrrhizin	*Klebsiella pneumoniae*
Alian, El-Ashwah, and Eid (1983)	*Glycyrrhiza glabra*	Glycyrrhizin	*Bacillus subtilis*
Yamashiki et al. (1999)	*Glycyrrhiza glabra*	Glycyrrhizin	Flaviviridae
Solis (1969)	*Abrus precatorius*	Glycyrrhizin	*Bacillus pumilus*

TABLE 5.76
Interactions of Grandiflorol with Microorganisms

Reference	Plant Name	Phytochemical	Organism
Solis (1969)	*Sesbania grandiflora*	Grandiflorol	*Bacillus pumilus*

TABLE 5.77
Interactions of Hydrocyanic Acid with Microorganisms

Reference	Plant Name	Phytochemical	Organism
Solis (1969)	*Erythrina variegata* var. *orientalis*	Hydrocyanic acid	*Escherichia coli*

TABLE 5.78
Interactions of Hydroxycitric Acid with Microorganisms

Reference	Plant Name	Phytochemical	Organism
De, De, and Banerjiee (1999)	*Garcinia cambogia*	Hydroxycitric acid	*Bacillus subtilis*
De, De, and Banerjiee (1999)	*Garcinia cambogia*	Hydroxycitric acid	*Escherichia coli*
De, De, and Banerjiee (1999)	*Garcinia cambogia*	Hydroxycitric acid	*Saccharomyces cerevisiae*

TABLE 5.79
Interactions of Karkadé Extracts with Microorganisms

Reference	Plant Name	Phytochemical	Organism
Alian, El-Ashwah, and Eid (1983)	*Hibiscus sabdariffa*	Karkadé extracts	*Escherichia coli*
Alian, El-Ashwah, and Eid (1983)	*Hibiscus sabdariffa*	Karkadé extracts	*Pseudomonas pyocyaneus*
Alian, El-Ashwah, and Eid (1983)	*Hibiscus sabdariffa*	Karkadé extracts	*Salmonella typhi*
Alian, El-Ashwah, and Eid (1983)	*Hibiscus sabdariffa*	Karkadé extracts	*Klebsiella pneumoniae*
Alian, El-Ashwah, and Eid (1983)	*Hibiscus sabdariffa*	Karkadé extracts	*Bacillus subtilis*

TABLE 5.80
Interactions of Kinotannic Acid with Microorganisms

Reference	Plant Name	Phytochemical	Organism
Solis (1969)	*Pterocarpus indicus*	Kinotannic acid	*Bacillus pumilus*

TABLE 5.81
Interactions of Licoflavanone with Microorganisms

Reference	Plant Name	Phytochemical	Organism
Fukui, Goto, and Tabata (1988)	*Glycyrrhiza glabra*	Licoflavanone	*Bacillus aubtilis*
Fukui, Goto, and Tabata (1988)	*Glycyrrhiza glabra*	Licoflavanone	*Staphylococcus aureus*
Fukui, Goto, and Tabata (1988)	*Glycyrrhiza glabra*	Licoflavanone	*Candida albicans*

TABLE 5.82
Interactions of Limonene with Microorganisms

Reference	Plant Name	Phytochemical	Organism
Ruberto et al. (2000)	*Crithmum maritimum*	Limonene	*Clostridium perfringens*
Inouye et al. (2000)	*Citrus junos*	Limonene	*Aspergillus fumigatus*
Singh et al. (2005)	*Cuminum cyminum*	Limonene	*Salmonella typhi*
Kubo, Himejima, and Muroi (1981)	*Elettaria cardamomum*	Limonene	*Saccharomyces cerevisiae*
Kubo, Himejima, and Muroi (1981)	*Elettaria cardamomum*	Limonene	*Penicillium chrysogenum*
De, De, and Banerjiee (1999)	*Trachyspermum ammi*	Limonene	*Saccharomyces cerevisiae*
De, De, and Banerjiee (1999)	*Myristica fragrans*	Limonene	*Bacillus subtilis*
De, De, and Banerjiee (1999)	*Carum carvi*	Limonene	*Escherichia coli*
Dorman and Deans (2000)	*Piper nigrum*	Limonene	*Brevibacterium linens*
Dorman and Deans (2000)	*Syzygium aromaticum*	Limonene	*Brevibacterium linens*
Dorman and Deans (2000)	*Pelargonium graveolens*	Limonene	*Brevibacterium linens*
Dorman and Deans (2000)	*Origanum vulgare*	Limonene	*Brevibacterium linens*
Dorman and Deans (2000)	*Thymus vulgaris*	Limonene	*Brevibacterium linens*
Jafri et al. (2001)	*Amomum subulatum*	Limonene	*Helicobacter pylori*
Inouye, Takizawa, and Yamaguchi (2001)	*Perilla frutescens*	Limonene	*Haemophilus influenzae*
Inouye, Takizawa, and Yamaguchi (2001)	*Perilla frutescens*	Limonene	*Streptococcus pyogenes*
Inouye, Takizawa, and Yamaguchi (2001)	*Perilla frutescens*	Limonene	*Streptococcus pneumoniae*
Inouye, Takizawa, and Yamaguchi (2001)	*Perilla frutescens*	Limonene	*Staphylococcus aureus*
Inouye, Takizawa, and Yamaguchi (2001)	*Perilla frutescens*	Limonene	*Escherichia coli*
Inouye, Takizawa, and Yamaguchi (2001)	*Thymus vulgaris*	Limonene	*Haemophilus influenzae*

(Continued)

TABLE 5.82 (*Continued*)
Interactions of Limonene with Microorganisms

Reference	Plant Name	Phytochemical	Organism
Inouye, Takizawa, and Yamaguchi (2001)	*Thymus vulgaris*	Limonene	*Streptococcus pyogenes*
Inouye, Takizawa, and Yamaguchi (2001)	*Thymus vulgaris*	Limonene	*Streptococcus pneumoniae*
Inouye, Takizawa, and Yamaguchi (2001)	*Thymus vulgaris*	Limonene	*Staphylococcus aureus*
Inouye, Takizawa, and Yamaguchi (2001)	*Thymus vulgaris*	Limonene	*Escherichia coli*
Inouye, Takizawa, and Yamaguchi (2001)	*Citrus medica*	Limonene	*Haemophilus influenzae*
Inouye, Takizawa, and Yamaguchi (2001)	*Citrus medica*	Limonene	*Streptococcus pyogenes*
Inouye, Takizawa, and Yamaguchi (2001)	*Citrus medica*	Limonene	*Streptococcus pneumoniae*
Inouye, Takizawa, and Yamaguchi (2001)	*Citrus medica*	Limonene	*Staphylococcus aureus*
Inouye, Takizawa, and Yamaguchi (2001)	*Citrus medica*	Limonene	*Escherichia coli*

TABLE 5.83
Interactions of Linalool with Microorganisms

Reference	Plant Name	Phytochemical	Organism
Lachowicz et al. (1998)	*Ocimum basilicum*	Linalool	*Lactobacillus curvatus*
Piccaglia et al. (1993)	*Lavandula angustifolia*	Linalool	*Micrococcus luteus*
Inouye et al. (2000)	*Lavandula*	Linalool	*Aspergillus fumigatus*
Kubo, Himejima, and Muroi (1981)	*Elettaria cardamomum*	Linalool	*Proteus vulgaris*
Kubo, Himejima, and Muroi (1981)	*Elettaria cardamomum*	Linalool	*Trichophyton mentagrophytes*
De, De, and Banerjiee (1999)	*Cinnamomum tamala*	Linalool	*Bacillus subtilis*
Elgayyar et al. (2001)	*Ocimum basilicum*	Linalool	*Aspergillus niger*
Elgayyar et al. (2001)	*Ocimum basilicum*	Linalool	Rhodotorula
Elgayyar et al. (2001)	*Ocimum basilicum*	Linalool	*Geotrichum candidum*
Elgayyar et al. (2001)	*Coriandrum sativum*	Linalool	*Aspergillus niger*
Elgayyar et al. (2001)	*Coriandrum sativum*	Linalool	Rhodotorula
Elgayyar et al. (2001)	*Coriandrum sativum*	Linalool	*Geotrichum candidum*
Dorman and Deans (2000)	*Piper nigrum*	Linalool	*Beneckea natriegens*

(*Continued*)

TABLE 5.83 (*Continued*)
Interactions of Linalool with Microorganisms

Reference	Plant Name	Phytochemical	Organism
Dorman and Deans (2000)	*Piper nigrum*	Linalool	*Serratia marcescens*
Dorman and Deans (2000)	*Syzygium aromaticum*	Linalool	*Beneckea natriegens*
Dorman and Deans (2000)	*Syzygium aromaticum*	Linalool	*Serratia marcescens*
Dorman and Deans (2000)	*Pelargonium graveolens*	Linalool	*Beneckea natriegens*
Dorman and Deans (2000)	*Pelargonium graveolens*	Linalool	*Serratia marcescens*
Dorman and Deans (2000)	*Myristica fragrans*	Linalool	*Brevibacterium linens*
Dorman and Deans (2000)	*Origanum vulgare*	Linalool	*Beneckea natriegens*
Dorman and Deans (2000)	*Origanum vulgare*	Linalool	*Serratia marcescens*
Dorman and Deans (2000)	*Thymus vulgaris*	Linalool	*Beneckea natriegens*
Dorman and Deans (2000)	*Thymus vulgaris*	Linalool	*Serratia marcescens*
Inouye, Takizawa, and Yamaguchi (2001)	*Coriandrum sativum*	Linalool	*Haemophilus influenzae*
Inouye, Takizawa, and Yamaguchi (2001)	*Coriandrum sativum*	Linalool	*Streptococcus pyogenes*
Inouye, Takizawa, and Yamaguchi (2001)	*Coriandrum sativum*	Linalool	*Streptococcus pneumoniae*
Inouye, Takizawa, and Yamaguchi (2001)	*Coriandrum sativum*	Linalool	*Staphylococcus aureus*
Inouye, Takizawa, and Yamaguchi (2001)	*Coriandrum sativum*	Linalool	*Escherichia coli*
Inouye, Takizawa, and Yamaguchi (2001)	*Lavandula latifolia*	Linalool	*Haemophilus influenzae*
Inouye, Takizawa, and Yamaguchi (2001)	*Lavandula latifolia*	Linalool	*Streptococcus pyogenes*
Inouye, Takizawa, and Yamaguchi (2001)	*Lavandula latifolia*	Linalool	*Streptococcus pneumoniae*
Inouye, Takizawa, and Yamaguchi (2001)	*Lavandula latifolia*	Linalool	*Staphylococcus aureus*
Inouye, Takizawa, and Yamaguchi (2001)	*Lavandula latifolia*	Linalool	*Escherichia coli*
Inouye, Takizawa, and Yamaguchi (2001)	*Lavandula angustifolia*	Linalool	*Haemophilus influenzae*
Inouye, Takizawa, and Yamaguchi (2001)	*Lavandula angustifolia*	Linalool	*Streptococcus pyogenes*
Inouye, Takizawa, and Yamaguchi (2001)	*Lavandula angustifolia*	Linalool	*Streptococcus pneumoniae*
Inouye, Takizawa, and Yamaguchi (2001)	*Lavandula angustifolia*	Linalool	*Staphylococcus aureus*
Inouye, Takizawa, and Yamaguchi (2001)	*Lavandula angustifolia*	Linalool	*Escherichia coli*

TABLE 5.84
Interactions of Linalyl Acetate with Microorganisms

Reference	Plant Name	Phytochemical	Organism
Piccaglia et al. (1993)	*Lavandula angustifolia*	Linalyl acetate	*Acinetobacter calcoaceticus*
Piccaglia et al. (1993)	*Lavandula angustifolia*	Linalyl acetate	*Alcaligenes faecalis*
Piccaglia et al. (1993)	*Lavandula angustifolia*	Linalyl acetate	*Bacillus subtilis*
Inouye et al. (2000)	Lavandula	Linalyl acetate	*Aspergillus fumigatus*
Kubo, Himejima, and Muroi (1981)	*Elettaria cardamomum*	Linalyl acetate	*Propionibacterium acnes*
Inouye, Takizawa, and Yamaguchi (2001)	*Lavandula angustifolia*	Linalyl acetate	*Haemophilus influenzae*
Inouye, Takizawa, and Yamaguchi (2001)	*Lavandula angustifolia*	Linalyl acetate	*Streptococcus pyogenes*
Inouye, Takizawa, and Yamaguchi (2001)	*Lavandula angustifolia*	Linalyl acetate	*Streptococcus pneumoniae*
Inouye, Takizawa, and Yamaguchi (2001)	*Lavandula angustifolia*	Linalyl acetate	*Staphylococcus aureus*
Inouye, Takizawa, and Yamaguchi (2001)	*Lavandula angustifolia*	Linalyl acetate	*Escherichia coli*

TABLE 5.85
Interactions of Linoleic Acid with Microorganisms

Reference	Plant Name	Phytochemical	Organism
Singh et al. (2005)	*Cuminum cyminum*	Linoleic acid	*Pseudomonas aeruginosa*
Solis (1969)	*Cassia tora*	Linoleic acid	*Sarcina lutea*

TABLE 5.86
Interactions of Luffaculin with Microorganisms

Reference	Plant Name	Phytochemical	Organism
Mishra, and Dubey (1990)	*Luffa acutangula*	Luffaculin	Human immunodeficiency virus

TABLE 5.87
Interactions of Luffin-a with Microorganisms

Reference	Plant Name	Phytochemical	Organism
Mishra, and Dubey (1990)	*Luffa cylindrica*	Luffin-a	Human immunodeficiency virus

TABLE 5.88
Interactions of Luffin-b with Microorganisms

Reference	Plant Name	Phytochemical	Organism
Mishra, and Dubey (1990)	*Luffa cylindrica*	Luffin-b	Human immunodeficiency virus

TABLE 5.89
Interactions of Luxetin with Microorganisms

Reference	Plant Name	Phytochemical	Organism
Solis (1969)	*Tamarindus indica*	Luxetin	*Sarcina lutea*

TABLE 5.90
Interactions of Malic Acid with Microorganisms

Reference	Plant Name	Phytochemical	Organism
Solis (1969)	*Tamarindus indica*	Malic acid	*Candida albicans*

TABLE 5.91
Interactions of Menthone with Microorganisms

Reference	Plant Name	Phytochemical	Organism
Dorman and Deans (2000)	*Piper nigrum*	Menthone	*Bacillus subtilis*
Dorman and Deans (2000)	*Syzygium aromaticum*	Menthone	*Bacillus subtilis*
Dorman and Deans (2000)	*Pelargonium graveolens*	Menthone	*Bacillus subtilis*
Dorman and Deans (2000)	*Myristica fragrans*	Menthone	*Flavobacterium suaveolens*
Dorman and Deans (2000)	*Origanum vulgare*	Menthone	*Bacillus subtilis*
Dorman and Deans (2000)	*Thymus vulgaris*	Menthone	*Bacillus subtilis*
Inouye, Takizawa, and Yamaguchi (2001)	*Mentha piperita*	Menthone	*Haemophilus influenzae*
Inouye, Takizawa, and Yamaguchi (2001)	*Mentha piperita*	Menthone	*Streptococcus pyogenes*
Inouye, Takizawa, and Yamaguchi (2001)	*Mentha piperita*	Menthone	*Streptococcus pneumoniae*
Inouye, Takizawa, and Yamaguchi (2001)	*Mentha piperita*	Menthone	*Staphylococcus aureus*
Inouye, Takizawa, and Yamaguchi (2001)	*Mentha piperita*	Menthone	*Escherichia coli*

TABLE 5.92
Interactions of Methanol Extracts with Microorganisms

Reference	Plant Name	Phytochemical	Organism
Cal and Wu (1996)	*Syzygium aromaticum*	Methanol extracts	*Streptococcus mutans*
Cal and Wu (1996)	*Syzygium aromaticum*	Methanol extracts	*Actinomyces viscosus*
Cal and Wu (1996)	*Syzygium aromaticum*	Methanol extracts	*Porphyromonas gingivalis*
Cal and Wu (1996)	*Syzygium aromaticum*	Methanol extracts	*Prevotella intermedia*

TABLE 5.93
Interactions of Methyl Chavicol with Microorganisms

Reference	Plant Name	Phytochemical	Organism
Lachowicz et al. (1998)	*Ocimum basilicum*	Methyl chavicol	*Saccharomyces cerevisiae*

TABLE 5.94
Interactions of Methyleugenol with Microorganisms

Reference	Plant Name	Phytochemical	Organism
Singh et al. (2005)	*Myristica fragrans*	Methyleugenol	*Staphylococcus aureus*
Kubo, Himejima, and Muroi (1981)	*Elettaria cardamomum*	Methyleugenol	*Enterobacter aerogenes*
Kubo, Himejima, and Muroi (1981)	*Elettaria cardamomum*	Methyleugenol	*Plasmodium ovale*

TABLE 5.95
Interactions of Momorcochin with Microorganisms

Reference	Plant Name	Phytochemical	Organism
Mishra, and Dubey (1990)	*Momordica cochinchinensis*	Momorcochin	Human immunodeficiency virus

TABLE 5.96
Interactions of Myrcene with Microorganisms

Reference	Plant Name	Phytochemical	Organism
Dorman and Deans (2000)	*Piper nigrum*	Myrcene	*Moraxella* sp.
Dorman and Deans (2000)	*Syzygium aromaticum*	Myrcene	*Moraxella* sp.
Dorman and Deans (2000)	*Myristica fragrans*	Myrcene	*Yersinia enterocolitica*
Dorman and Deans (2000)	*Origanum vulgare*	Myrcene	*Moraxella* sp.
Dorman and Deans (2000)	*Thymus vulgaris*	Myrcene	*Moraxella* sp.
Jafri et al. (2001)	*Amomum subulatum*	Myrcene	*Helicobacter pylori*

TABLE 5.97
Interactions of Neral with Microorganisms

Reference	Plant Name	Phytochemical	Organism
Inouye, Takizawa, and Yamaguchi (2001)	*Cymbopogon*	Neral	*Haemophilus influenzae*
Inouye, Takizawa, and Yamaguchi (2001)	*Cymbopogon*	Neral	*Streptococcus pyogenes*
Inouye, Takizawa, and Yamaguchi (2001)	*Cymbopogon*	Neral	*Streptococcus pneumoniae*
Inouye, Takizawa, and Yamaguchi (2001)	*Cymbopogon*	Neral	*Staphylococcus aureus*
Inouye, Takizawa, and Yamaguchi (2001)	*Cymbopogon*	Neral	*Escherichia coli*

TABLE 5.98
Interactions of Nerol with Microorganisms

Reference	Plant Name	Phytochemical	Organism
Dorman and Deans (2000)	*Piper nigrum*	Nerol	*Aeromonas hydrophila*
Dorman and Deans (2000)	*Syzygium aromaticum*	Nerol	*Aeromonas hydrophila*
Dorman and Deans (2000)	*Myristica fragrans*	Nerol	*Erwinia carotovora*
Dorman and Deans (2000)	*Origanum vulgare*	Nerol	*Aeromonas hydrophila*
Dorman and Deans (2000)	*Thymus vulgaris*	Nerol	*Aeromonas hydrophila*

TABLE 5.99
Interactions of Oxymethylanthraquinone with Microorganisms

Reference	Plant Name	Phytochemical	Organism
Solis (1969)	*Cassia aata*	Oxymethylanthraquinone	*Escherichia coli*
Solis (1969)	*Cassia fistula*	Oxymethylanthraquinone	*Escherichia coli*

TABLE 5.100
Interactions of Pachyrrhizid with Microorganisms

Reference	Plant Name	Phytochemical	Organism
Solis (1969)	*Pachyrhizus erosus*	Pachyrrhizid	*Sarcina lutea*

TABLE 5.101
Interactions of PAP-S with Microorganisms

Reference	Plant Name	Phytochemical	Organism
Tomasi et al. (1982)	*Phytolacca americana*	PAP-S	Poliovirus

TABLE 5.102
Interactions of Paracymene with Microorganisms

Reference	Plant Name	Phytochemical	Organism
Piccaglia et al. (1993)	*Thymus vulgaris*	Paracymene	*Brevibacterium linens*
Piccaglia et al. (1993)	*Satureja montana*	Paracymene	*Brevibacterium linens*
Singh et al. (2005)	*Nigella sativa*	Paracymene	*Staphylococcus aureus*
Singh et al. (2005)	*Cuminum cyminum*	Paracymene	*Bacillus subtilis*
De, De, and Banerjiee (1999)	*Cuminum cyminum*	Paracymene	*Bacillus subtilis*
Dorman and Deans (2000)	*Piper nigrum*	Paracymene	*Citrobacter freundii*
Dorman and Deans (2000)	*Syzygium aromaticum*	Paracymene	*Citrobacter freundii*
Dorman and Deans (2000)	*Pelargonium graveolens*	Paracymene	*Citrobacter freundii*
Dorman and Deans (2000)	*Origanum vulgare*	Paracymene	*Citrobacter freundii*
Dorman and Deans (2000)	*Thymus vulgaris*	Paracymene	*Citrobacter freundii*

TABLE 5.103
Interactions of Paramenthone with Microorganisms

Reference	Plant Name	Phytochemical	Organism
Inouye, Takizawa, and Yamaguchi (2001)	*Mentha piperita*	Paramenthone	*Haemophilus influenzae*
Inouye, Takizawa, and Yamaguchi (2001)	*Mentha piperita*	Paramenthone	*Streptococcus pyogenes*
Inouye, Takizawa, and Yamaguchi (2001)	*Mentha piperita*	Paramenthone	*Streptococcus pneumoniae*
Inouye, Takizawa, and Yamaguchi (2001)	*Mentha piperita*	Paramenthone	*Staphylococcus aureus*
Inouye, Takizawa, and Yamaguchi (2001)	*Mentha piperita*	Paramenthone	*Escherichia coli*

TABLE 5.104
Interactions of Peperine with Microorganisms

Reference	Plant Name	Phytochemical	Organism
De, De, and Banerjiee (1999)	*Piper nigrum*	Peperine	*Bacillus subtilis*
De, De, and Banerjiee (1999)	*Piper nigrum*	Peperine	*Escherichia coli*
De, De, and Banerjiee (1999)	*Piper nigrum*	Peperine	*Saccharomyces cerevisiae*

TABLE 5.105
Interactions of Perillaldehyde with Microorganisms

Reference	Plant Name	Phytochemical	Organism
Inouye et al. (2000)	*Perilla frutescens*	Perillaldehyde	*Aspergillus fumigatus*
Inouye, Takizawa, and Yamaguchi (2001)	*Perilla frutescens*	Perillaldehyde	*Haemophilus influenzae*
Inouye, Takizawa, and Yamaguchi (2001)	*Perilla frutescens*	Perillaldehyde	*Streptococcus pyogenes*
Inouye, Takizawa, and Yamaguchi (2001)	*Perilla frutescens*	Perillaldehyde	*Streptococcus pneumoniae*
Inouye, Takizawa, and Yamaguchi (2001)	*Perilla frutescens*	Perillaldehyde	*Staphylococcus aureus*
Inouye, Takizawa, and Yamaguchi (2001)	*Perilla frutescens*	Perillaldehyde	*Escherichia coli*

TABLE 5.106
Interactions of Phenolic Glycosides with Microorganisms

Reference	Plant Name	Phytochemical	Organism
Spiliotis et al. (1998)	*Moringa oleifera*	Phenolic glycosides	*Bacillus cereus*
Spiliotis et al. (1998)	*Moringa oleifera*	Phenolic glycosides	*Candida albicans*
Spiliotis et al. (1998)	*Moringa oleifera*	Phenolic glycosides	*Streptococcus faecalis*
Spiliotis et al. (1998)	*Moringa oleifera*	Phenolic glycosides	*Staphylococcus aureus*
Spiliotis et al. (1998)	*Moringa oleifera*	Phenolic glycosides	*Staphylococcus epidermidis*
Spiliotis et al. (1998)	*Moringa oleifera*	Phenolic glycosides	*Bacillus subtilis*
Spiliotis et al. (1998)	*Moringa oleifera*	Phenolic glycosides	*Pseudomonas aeruginosa*
Spiliotis et al. (1998)	*Moringa oleifera*	Phenolic glycosides	*Escherichia coli*
Spiliotis et al. (1998)	*Moringa oleifera*	Phenolic glycosides	*Aspergillus niger*
De, De, and Banerjiee (1999)	*Allium cepa*	Phenolic glycosides	*Bacillus subtilis*
De, De, and Banerjiee (1999)	*Allium cepa*	Phenolic glycosides	*Escherichia coli*
De, De, and Banerjiee (1999)	*Allium cepa*	Phenolic glycosides	*Saccharomyces cerevisiae*
De, De, and Banerjiee (1999)	*Punica granatum*	Phenolic glycosides	*Saccharomyces cerevisiae*
Lans and Brown (1998)	*Bryophyllum pinnatum*	Phenolic glycosides	To promote chicken heath
Solis (1969)	*Clitoria ternatea*	Phenolic glycosides	*Bacillus pumilus*

TABLE 5.107
Interactions of Phytosterols with Microorganisms

Reference	Plant Name	Phytochemical	Organism
Solis (1969)	*Caesalpinia crista*	Phytosterols	*Candida albicans*

TABLE 5.108
Interactions of Pinocembrin with Microorganisms

Reference	Plant Name	Phytochemical	Organism
Fukui, Goto, and Tabata (1988)	*Glycyrrhiza glabra*	Pinocembrin	*Bacillus subtilis*
Fukui, Goto, and Tabata (1988)	*Glycyrrhiza glabra*	Pinocembrin	*Staphylococcus aureus*
Fukui, Goto, and Tabata (1988)	*Glycyrrhiza glabra*	Pinocembrin	*Candida albicans*

TABLE 5.109
Interactions of Piperine for Lipid Peroxidation

Reference	Plant Name	Phytochemical	Activity
Vijayakumar, Surya, and Nalini (2004)	*Piper nigrum*	Piperine	Lipid peroxidation in rats

TABLE 5.110
Interactions of Pithecolobin with Microorganisms

Reference	Plant Name	Phytochemical	Organism
Solis (1969)	*Samanea saman*	Pithecolobin	*Sarcina lutea*

TABLE 5.111
Interactions of Prenylated Xanthones with Microorganisms

Reference	Plant Name	Phytochemical	Organism
Suksmrarnh et al. (2003)	*Garcinia mangostana*	Prenylated xanthones	*Mycobacterium tuberculosis*

TABLE 5.112
Interactions of Pterygospermin with Microorganisms

Reference	Plant Name	Phytochemical	Organism
Das, Kurup, and Narasimha Rao (1957)	*Moringa pterygosperma*	Pterygospermin	*Escherichia coli*
Caceres et al. (1991)	*Moringa oleifera*	Pterygospermin	*Pseudomonas aeruginosa*
Caceres et al. (1991)	*Moringa oleifera*	Pterygospermin	*Staphylococcus aureus*

TABLE 5.113
Interactions of Punicalagins with Microorganisms

Reference	Plant Name	Phytochemical	Organism
De, De, and Banerjiee (1999)	*Punica granatum*	Punicalagins	*Escherichia coli*

TABLE 5.114
Interactions of Resins with Microorganisms

Reference	Plant Name	Phytochemical	Organism
Solis (1969)	*Caesalpinia pulcherrima*	Resins	*Candida albicans*
Solis (1969)	*Mimosa pudica*	Resins	*Sarcina lutea*

TABLE 5.115
Interactions of Rotenone with Microorganisms

Reference	Plant Name	Phytochemical	Organism
Solis (1969)	*Derris elliptica*	Rotenone	*Bacillus pumilus*

TABLE 5.116
Interactions of Sabinene with Microorganisms

Reference	Plant Name	Phytochemical	Organism
Singh et al. (2005)	*Myristica fragrans*	Sabinene	*Bacillus cereus*
Dorman and Deans (2000)	*Myristica fragrans*	Sabinene	*Proteus vulgaris*

TABLE 5.117
Interactions of Safrole with Microorganisms

Reference	Plant Name	Phytochemical	Organism
Singh et al. (2005)	*Myristica fragrans*	Safrole	*Escherichia coli*
Kubo, Himejima, and Muroi (1981)	*Elettaria cardamomum*	Safrole	*Escherichia coli*
De, De, and Banerjiee (1999)	*Myristica fragrans*	Safrole	*Escherichia coli*

TABLE 5.118
Interactions of Saponins with Microorganisms

Reference	Plant Name	Phytochemical	Organism
Solis (1969)	*Cassia tora*	Saponins	*Escherichia coli*
Solis (1969)	*Erythrina variegata* var. *orientalis*	Saponins	*Sarcina lutea*
Solis (1969)	*Mimosa pudica*	Saponins	*Bacillus pumilus*
Solis (1969)	*Tamarindus indica*	Saponins	*Sarcina lutea*

TABLE 5.119
Interactions of Seeds Extracts with Microorganisms

Reference	Plant Name	Phytochemical	Organism
Sahu and Chakrabarty (1994)	*Cassia tora*	Seeds extracts	*Bacillus anthracis*
Sahu and Chakrabarty (1994)	*Cassia tora*	Seeds extracts	*Bacillus pumilus*
Sahu and Chakrabarty (1994)	*Cassia tora*	Seeds extracts	*Salmonella paratyphi*
Sahu and Chakrabarty (1994)	*Cassia tora*	Seeds extracts	*Staphylococcus albus*
Sahu and Chakrabarty (1994)	*Cassia tora*	Seeds extracts	*Xanthomonas campestris*
Sahu and Chakrabarty (1994)	*Cassia tora*	Seeds extracts	*Xanthomonas malvacearum*
Singh et al. (2005)	*Piper nigrum*	Seeds extracts	*Staphylococcus aureus*
Singh et al. (2005)	*Piper nigrum*	Seeds extracts	*Pseudomonas aeruginosa*
Singh et al. (2005)	*Piper nigrum*	Seeds extracts	*Bacillus subtilis*
Singh et al. (2005)	*Piper nigrum*	Seeds extracts	*Bacillus cereus*

TABLE 5.120
Interactions of S-Methyl Cysteine Sulfoxide with Diabetic Oraganism

Reference	Plant Name	Phytochemical	Organism
Kumari, Mathew, and Augusti (1995)	*Allium cepa*	S-Methyl cysteine sulfoxide	Antidiabetic in rats

TABLE 5.121
Interactions of Sotolone as an Antioxidative Property

Reference	Plant Name	Phytochemical	Activity
Mansour and Khalil (2000)	*Trigonella foenum-graecum*	Sotolone	Antioxidant

TABLE 5.122
Interactions of Sterols with Microorganisms

Reference	Plant Name	Phytochemical	Organism
Chaurasia and Jain (1978)	*Pongamia pinnata*	Sterols	*Salmonella typhi*
Chaurasia and Jain (1978)	*Cinnamomum zeylainchem*	Sterols	*Staphylococcus aureus*
Chaurasia and Jain (1978)	*Anethum graveolens*	Sterols	*Bacillus mycoides*
Chaurasia and Jain (1978)	*Anethum gaveolens*	Sterols	*Staphylococcus aureus*
Chaurasia and Jain (1978)	*Azadirachta indica*	Sterols	*Bacillus mycoides*
Chaurasia and Jain (1978)	*Azadirachta indica*	Sterols	*Staphylococcus aureus*

TABLE 5.123
Interactions of Tamarind Extracts with Microorganisms

Reference	Plant Name	Phytochemical	Organism
Alian, El-Ashwah, and Eid (1983)	*Tamarindus indica*	Tamarind extracts	*Escherichia coli*
Alian, El-Ashwah, and Eid (1983)	*Tamarindus indica*	Tamarind extracts	*Pseudomonas pyocyaneus*
Alian, El-Ashwah, and Eid (1983)	*Tamarindus indica*	Tamarind extracts	*Salmonella typhi*
Alian, El-Ashwah, and Eid (1983)	*Tamarindus indica*	Tamarind extracts	*Klebsiella pneumoniae*
Alian, El-Ashwah, and Eid (1983)	*Tamarindus indica*	Tamarind extracts	*Bacillus subtilis*

TABLE 5.124
Interactions of Tannic Acid with Microorganisms

Reference	Plant Name	Phytochemical	Organism
Solis (1969)	*Caesalpinia sappan*	Tannic acid	*Bacillus pumilus*

TABLE 5.125
Interactions of Tannins with Microorganisms

Reference	Plant Name	Phytochemical	Organism
Hsieh, Mau, and Huang (2001)	*Cornus officinalis*	Tannins	*Bacillus subtilis*
Hsieh, Mau, and Huang (2001)	*Cornus officinalis*	Tannins	*Pseudomonas aeruginosa*
Hsieh, Mau, and Huang (2001)	*Cornus officinalis*	Tannins	*Debaryomyces hansenii*
Hsieh, Mau, and Huang (2001)	*Cornus officinalis*	Tannins	*Aspergillus niger*
Solis (1969)	*Acacia farnesiana*	Tannins	*Bacillus pumilus*
Solis (1969)	*Bauhinia malabarica*	Tannins	*Bacillus pumilus*
Solis (1969)	*Caesalpinia pulcherrima*	Tannins	*Bacillus pumilus*
Solis (1969)	*Cassia alata*	Tannins	*Sarcina lutea*
Solis (1969)	*Cassia fistula*	Tannins	*Bacillus pumilus*
Solis (1969)	*Mimosa pudica*	Tannins	*Candida albicans*
Solis (1969)	*Pithecellobium dulce*	Tannins	*Sarcina lutea*
Solis (1969)	*Sesbania grandiflora*	Tannins	*Sarcina lutea*
Solis (1969)	*Tamarindus indica*	Tannins	*Bacillus pumilus*

TABLE 5.126
Interactions of Tartaric Acid with Microorganisms

Reference	Plant Name	Phytochemical	Organism
Solis (1969)	*Parkia javanica*	Tartaric acid	*Sarcina lutea*
Solis (1969)	*Tamarindus indica*	Tartaric acid	*Escherichia coli*

TABLE 5.127
Interactions of Tau-Terpinen with Microorganisms

Reference	Plant Name	Phytochemical	Organism
Piccaglia et al. (1993)	*Thymus vulgaris*	Tau-terpinen	*Brochothrix thermosphacta*

TABLE 5.128
Interactions of Taxicarol with Microorganisms

Reference	Plant Name	Phytochemical	Organism
Solis (1969)	*Derris elliptica*	Taxicarol	*Candida albicans*

TABLE 5.129
Interactions of Tephrosin with Microorganisms

Reference	Plant Name	Phytochemical	Organism
Solis (1969)	*Derris elliptica*	Tephrosin	*Escherichia coli*

TABLE 5.130
Interactions of Terpenes with Microorganisms

Reference	Plant Name	Phytochemical	Organism
Chaurasia and Jain (1978)	*Pongamia pinnata*	Terpenes	*Bacillus anthracis*
Chaurasia and Jain (1978)	*Pongamia pinnata*	Terpenes	*Pseudomonas mangiferae-indicae*
Chaurasia and Jain (1978)	*Cinnamomum zeylainchem*	Terpenes	*Bacillus mycoides*
Chaurasia and Jain (1978)	*Anethum graveolens*	Terpenes	*Bacillus anthracis*
Chaurasia and Jain (1978)	*Anethum graveolens*	Terpenes	*Pseudomonas mangiferae-indicae*
Chaurasia and Jain (1978)	*Anethum graveolens*	Terpenes	*Salmonella typhi*
Chaurasia and Jain (1978)	*Azadirachta indica*	Terpenes	*Bacillus anthracis*
Chaurasia and Jain (1978)	*Azadirachta indica*	Terpenes	*Pseudomonas mangiferae-indicae*
Chaurasia and Jain (1978)	*Azadirachta indica*	Terpenes	*Salmonella typhi*
De, De, and Banerjiee (1999)	*Carum carvi*	Terpenes	*Saccharomyces cerevisiae*

TABLE 5.131
Interactions of Terpinen-4-ol with Microorganisms

Reference	Plant Name	Phytochemical	Organism
Inouye et al. (2000)	*Melaleuca alternifolia*	Terpinen-4-ol	*Aspergillus fumigatus*
Singh et al. (2005)	*Myristica fragrans*	Terpinen-4-ol	*Bacillus subtilis*
Dorman and Deans (2000)	*Myristica fragrans*	Terpinen-4-ol	*Bacillus subtilis*
Inouye, Takizawa, and Yamaguchi (2001)	*Melaleuca alternifolia*	Terpinen-4-ol	*Haemophilus influenzae*
Inouye, Takizawa, and Yamaguchi (2001)	*Melaleuca alternifolia*	Terpinen-4-ol	*Streptococcus pyogenes*
Inouye, Takizawa, and Yamaguchi (2001)	*Melaleuca alternifolia*	Terpinen-4-ol	*Streptococcus pneumoniae*
Inouye, Takizawa, and Yamaguchi (2001)	*Melaleuca alternifolia*	Terpinen-4-ol	*Staphylococcus aureus*
Inouye, Takizawa, and Yamaguchi (2001)	*Melaleuca alternifolia*	Terpinen-4-ol	*Escherichia coli*

TABLE 5.132
Interactions of Thujone with Microorganisms

Reference	Plant Name	Phytochemical	Organism
Dorman and Deans (2000)	*Myristica fragrans*	Thujone	*Brochothrix thermosphacta*

TABLE 5.133
Interactions of Thymol with Microorganisms

Reference	Plant Name	Phytochemical	Organism
Piccaglia et al. (1993)	*Thymus vulgaris*	Thymol	*Lactobacillus plantarum*
Piccaglia et al. (1993)	*Thymus vulgaris*	Thymol	*Leuconostoc cremoris*
Piccaglia et al. (1993)	*Thymus vulgaris*	Thymol	*Acinetobacter calcoaceticus*
Singh et al. (2005)	*Nigella sativa*	Thymol	*Salmonella typhi*
De, De, and Banerjiee (1999)	*Trachyspermum ammi*	Thymol	*Bacillus subtilis*
Elgayyar et al. (2001)	*Origanum vulgare*	Thymol	*Salmonella typhimurium*
Elgayyar et al. (2001)	*Origanum vulgare*	Thymol	*Yersinia enterocolitica*
Elgayyar et al. (2001)	*Origanum vulgare*	Thymol	*Escherichia coli*
Elgayyar et al. (2001)	*Origanum vulgare*	Thymol	*Staphylococcus aureus*
Dorman and Deans (2000)	*Piper nigrum*	Thymol	*Salmonella pullorum*
Dorman and Deans (2000)	*Syzygium aromaticum*	Thymol	*Salmonella pullorum*
Dorman and Deans (2000)	*Pelargonium graveolens*	Thymol	*Salmonella pullorum*
Dorman and Deans (2000)	*Myristica fragrans*	Thymol	*Acinetobacter calcoaceticus*
Dorman and Deans (2000)	*Origanum vulgare*	Thymol	*Salmonella pullorum*
Dorman and Deans (2000)	*Thymus vulgaris*	Thymol	*Salmonella pullorum*
Inouye, Takizawa, and Yamaguchi (2001)	*Thymus vulgaris*	Thymol	*Haemophilus influenzae*
Inouye, Takizawa, and Yamaguchi (2001)	*Thymus vulgaris*	Thymol	*Streptococcus pyogenes*
Inouye, Takizawa, and Yamaguchi (2001)	*Thymus vulgaris*	Thymol	*Streptococcus pneumoniae*
Inouye, Takizawa, and Yamaguchi (2001)	*Thymus vulgaris*	Thymol	*Staphylococcus aureus*
Inouye, Takizawa, and Yamaguchi (2001)	*Thymus vulgaris*	Thymol	*Escherichia coli*

TABLE 5.134
Interactions of Thymol Methyl Ether with Microorganisms

Reference	Plant Name	Phytochemical	Organism
Ruberto et al. (2000)	*Crithmum maritimum*	Thymol methyl ether	*Clostridium perfringens*

TABLE 5.135
Interactions of Thymoquinone with Microorganisms

Reference	Plant Name	Phytochemical	Organism
Singh et al. (2005)	*Nigella sativa*	Thymoquinone	*Bacillus cereus*
De, De, and Banerjiee (1999)	*Nigella sativa*	Thymoquinone	*Bacillus subtilis*
De, De, and Banerjiee (1999)	*Nigella sativa*	Thymoquinone	*Escherichia coli*
De, De, and Banerjiee (1999)	*Nigella sativa*	Thymoquinone	*Saccharomyces cerevisiae*

TABLE 5.136
Interactions of Toxalbumin Abrin with Microorganisms

Reference	Plant Name	Phytochemical	Organism
Solis (1969)	*Abrus precatorius*	Toxalbumin abrin	*Escherichia coli*

TABLE 5.137
Interactions of Trichosanthin with Microorganisms

Reference	Plant Name	Phytochemical	Organism
Zheng, Ben, and Jin (1999)	*Trichosanthes kirilowii*	Trichosanthin	Human immunodeficiency virus-1
Mishra, and Dubey (1990)	*Trichosanthes kirilowii*	Trichosanthin	Human immunodeficiency virus

TABLE 5.138
Interactions of Vitexin with Microorganisms

Reference	Plant Name	Phytochemical	Organism
Solis (1969)	*Tamarindus indica*	Vitexin	*Bacillus pumilus*

TABLE 5.139
Interactions of Xanthones with Microorganisms

Reference	Plant Name	Phytochemical	Organism
Gopalakrishnan, Banumathi, and Suresh (1997)	*Garcinia mangostana*	Xanthones	*Fusarium oxysporum vasinfectum*
Gopalakrishnan, Banumathi, and Suresh (1997)	*Garcinia mangostana*	Xanthones	*Alternaria tenuis*
Gopalakrishnan, Banumathi, and Suresh (1997)	*Garcinia mangostana*	Xanthones	*Drechslera oryzae*

may protect the organisms against any environmental stress. In addition to their biofilm inhibitory activity, some of them induce the pathogenicity of the microorganism. For example, curcumin interacts with the defense pathways of *Salmonella enterica* serovar *Typhimurium* (*S. typhimurium*) to induce its pathogenicity in a murine model. In a study, Yamamoto et al. (1988) reported that curcumin downregulates *Salmonella* Pathogenicity Island 1 (SPI1) genes required for entry into epithelial cells and upregulates *Salmonella* Pathogenicity Island 2 (SPI2) genes required to intracellular survival. Both these genes are regulated by a common regulatory gene called the *PhoPQ* system. This article warns the use of curcumin during *Salmonella* outbreaks.

There are other bacteria such as *S. aureus*, *E. coli,* and *S. lutea* that are inhibited by curcumin because of its phototoxic effect through hydrogen peroxide (Dahl et al., 1989). *Bacillus subtilis* is inhibited by curcumin through FtsZ assembly modification because of discomposing its GTPase activity (Dahl et al., 1989). Curcumin inhibits the growth of *H. pylori* through the inhibition of shikimate dehydrogenase gene activity (Han et al., 2006).

In contrast, curcumin fails to induce the cell death in *S. typhimurium* TA1535/ pSK1002 and *E. coli* K-12 strains by inhibiting SOS gene induction and mutagenesis by UV light (Oda, 1995). Curcumin also protects *E. coli*, *B. megaterium*, and *B. pumilus* from gamma radiation–induced DNA inactivation (Sharma, Gautam, and Jadhav, 2000). These studies indicate that natural products such as curcumin, a strong antioxidative agent, may play the role of both the inhibition and induction of organismal growth depending on their stress, concentration, and species.

Another study from the natural products of volatile oils from plant exhibits the antimicrobial activities. Thymol was found to be inhibiting a widest range of microorganisms (Dorman, 1999). It is followed by carvacrol that inhibits fewer microorganisms than thymol, and the following order provides for other volatile oils on their range of antimicrobial activity on fewer organisms. The range of order of volatile oil includes α-terpineol; terpinen-4-ol; eugenol; (±)-linalool; (−)-thujone; δ-3-carene; *cis*-hex-3-an-1-ol; geranyl acetate; (*cis* + *trans*) citral; nerol; geraniol; menthone; β-pinene; *R*(+)-limonene; α-pinene; α-terpinene; borneol; (+)-sabinene; γ-terpinene; citronellal; terpinolene; 1,8-cineole; bornyl acetate; carvacrol methyl ether; myrcene; β-caryophyllene; α-bisabolol; α-phellandrene; α-humulene; β-ocimene; aromadendrene; and *p*-cymene. Because of its antimicrobial activities, these volatile oils are used in preserving the foods. The aromatic volatile oils prevent the loss of organoleptic properties, microbial contamination, and the onset of spoilage. This is due to the strong antioxidant property of the volatile oil that can inhibit the free radical–induced organoleptic degradation (Youdim and Deans, 1999).

5.4   CONCLUSION

Apart from the bactericidal activities of natural products, some of them also participate in protecting the organisms from oxidative stress. In most of the cases, the metabolites of natural products from the bacterial digestion may modulate the therapeutic activity and bioavailability of that compound. Therefore, caution should be exercised when a more potent drug screened by in vitro methods most probably failed its therapeutic activity during in vivo administration. This may be due to the drug–natural product–microbial interactions in the gut. In addition, the ecology of the gut may vary between genders, age, racial, and geographical presence. Also, unlike the traditional antibiotic therapy, where it kills the entire gut microbial flora, the natural products selectively kill the specific species providing additional information to address the novel antibiotic therapy in future. The recent development in biofilm and other human microbiota data warrant devising new drug designing strategies to address all these issues.

ACKNOWLEDGMENTS

The data analysis help was provided by Ria Ghosh and Dipal Parekh.

REFERENCES

Agans, R., L. Rigsbee, H. Kenche, S. Michail, H. J. Khamis, and O. Paliy. 2011. Distal gut microbiota of adolescent children is different from that of adults. *FEMS Microbiol Ecol* 77 (2):404–12.

Alian, A., E. El-Ashwah, and N. Eid. 1983. Antimicrobial properties of some Egyptian non-alcoholic beverages with special reference to tamarind. *Egypt J Food Sci* 11:109–14.

Barl, M. L., H. Kushunoki, H. Furukawa et al. 1999. Inhibition of growth of Escherichia coli O157:H7 in fresh radish (Raphanus sativus L.) sprout production by calcinated calcium. *J Food Prot* 62 (2):128–32.

Caceres, A., O. Cabrera, O. Morales et al. 1991. Pharmacological properties of Moringa oleifera. 1: preliminary screening for antimicrobial activity. *J Ethanopharmacaol* 33:213–6.

Caceres, A., A. Saravia, S. Rizzo et al. 1992. Pharmacologic properties of Moringa oleifera. 2: screening for antispasmodic, antiinflammatory and diuretic activity. *J Ethanopharmacaol* 36:233–7.

Cal, L., and C. Wu. 1996. Compounds from Syzygium aromaticum possessing growth inhibitory activity against oral pathogens. *J Nat Prod* 59:987–90.

Chaieb, K., B. Kouidhi, H. Jrah et al. 2011. Antibacterial activity of Thymoquinone, an active principle of Nigella sativa and its potency to prevent bacterial biofilm formation. *BMC Complement Altern Med* 11:29.

Chaurasia, S. C., and P. C. Jain. November–December 1978. Antibacterial activity of essential oils of four medicinal plants. *Indian J Hosp Pharm* 166–8.

Dahl, T. A., W. M. McGowan, M. A. Shand, and V. S. Srinivasan. 1989. Photokilling of bacteria by the natural dye curcumin. *Arch Microbiol* 151 (2):183–5.

Das, B. R., P. A. Kurup, and P. L. Narasimha Rao. 1957. Antibiotic principle from Moringa pterygosperma. VII. Antibacterial activity and chemical structure of compounds related to pterygospermin. *Indian J Med Res* 45 (2):191–6.

De, M., A. Krishna De, and A. B. Banerjiee. 1999. Antimicrobial screening of some Indian spices. *Phytother Res* 13(7):616–8.

Dorman, H. J. D. 1999. Phytochemistry and bioactive properties of plant volatile oils: antibacterial, antifungal and antioxidant activities. PhD Thesis, University of Strathclyde, Glasgow, Glasgow City.

Dorman, H. J., and S. G. Deans. 2000. Antimicrobial agents from plants: antibacterial activity of plant volatile oils. *J Appl Microbiol* 88 (2):308–16.

Eilert, U., B. Wolters, and A. Nahrstedt. 1981. The antibiotic principle of seeds of Moringa oleifera and Moringa stenopetala. *Planta Med* 42:55–61.

Elgayyar, M., F. A. Draughon, D. A. Golden et al. 2001. Antimicrobial activity of essential oils from plants against selected pathogenic and saprophytic microorganisms. *J Food Prot* 64 (7):1019–24.

Fukui, H., K. Goto, and M. Tabata. 1988. Two antimicrobial flavanones from the leaves of Glycyrrhiza glabra. *Chem Pharm Bull* 36 (10):4174–6.

Giron, L. M., V. Freire, A. Alonza et al. 1991. Ethnobotanical survey of the medicinal flora used by the Caribs of Guatemala. *J Ethanopharmacol* 34:171–87.

Gopalakrishnan, G., B. Banumathi, and G. Suresh. 1997. Evaluation of the antifungal activity of natural xanthones from Garcinia mangostana and their synthetic derivatives. *J Nat Prod* 60:519–24.

Graf, B. A., W. Mullen, S. T. Caldwell et al. 2005. Disposition and metabolism of [2-14C] quercetin-4'-glucoside in rats. *Drug Metab Dispos* 33 (7):1036–43.

Gruner, E., A. G. Steigerwalt, D. G. Hollis et al. 1994. Human infections caused by Brevibacterium casei, formerly CDC groups B-1 and B-3. *J Clin Microbiol* 32 (6):1511–8.

Guillen, M. D., and M. J. Manzanos. 1996. A study of several parts of the plant Foeniculum vulgare as a source of compounds with industrial interest. *Food Res Int* 29 (1):85–8.

Hammer, K. A., C. F. Carson, and T. V. Riley. 1999. Antimicrobial activity of essential oils and other plant extracts. *J Appl Microbiol* 86:985–90.

Han, C., L. Wang, K. Yu et al. 2006. Biochemical characterization and inhibitor discovery of shikimate dehydrogenase from Helicobacter pylori. *FEBS J* 273 (20):4682–92.

Hart, M., and P. S. Rathee. 1995. Antibacterial activity of the unsaponifiable fraction of the fixed oils of trichosanthes seeds. *Asian J Chem* 7 (4):909–11.

Hsieh, P. C., J. L. Mau, and A. H. Huang. 2001. Antimicrobial effect of various combinations of plant extracts. *Food Microbiol* 18:35–43.

Inouye, S., T. Takizawa, and H. Yamaguchi. 2001. Antibacterial activity of essential oils and their major constituents against respiratory tract pathogens by gaseous contact. *J Antimicrobiol Chemother* 47:565–73.

Inouye, S., T. Tsuruoka, M. Watanabe et al. 2000. Inhibitory effect of essential oils on apical growth of Aspergillus fumigatus by vapour contact. *Mycoses* 43 (1–2):17–23.

Jafri, M. A., J. Farah, K. Javed et al. (2001). Evaluation of the gastric antiulcerogenic effect of large cardamom (fruits of Amomum subulatum Roxb). *J Ethanopharmacol* 75:89–94.

Kendler, B. S. 1987. Garlic and onion: a review of their relationship to cardiovascular disease. *Prev Med* 16:670–85.

Kubo, I., M. Himejima, and H. Muroi. 1981. Antimicrobial activity of flavor components of cardamom Elettaria cardamomum (Dahl et al.) seed. *J Agric Food Chem* 39:1984–6.

Kumaresan, P. T., A. V. Kumar, P. R. Arivarasu, and S. Balasubramanian. 2001. Anti-bacterial activity of aerial parts of Alternanthera sessilis. *Indian J Nat Prod* 17:23–4.

Kumari, K., B. Mathew, and K. T. Augusti. 1995. Antidiabetic and hypolipidemic effects of S-methyl cysteine sulfoxide isolated from Allium cepa Linn. *Indian J Biochem Biophys* 32:49–54.

Kusamran, W., A. Tepsuwan, and P. Kupradinun. 1998. Antimutagenic and anticarcinogenic potentials of some Thai vegetables. *Mutat Res* 402:247–58.

Lachowicz, K. J., G. P. Jones, D. R. Briggs et al. 1998. The synergistic preservative effects of the essential oils of sweet basil (Ocimum basilicum L. against acid-tolerant food microflora. *Lett Appl Microbiol* 26 (3):209–14.

Lans, C., and G. Brown. 1998. Observations on ethnoveterinary medicines in Trinidad and Tobago. *Prev Vet Med* 35:125–42.

Latgé, J. P. 1999. Aspergillus fumigatus and aspergillosis. *Clin Microbiol Rev* 12 (2):310–50.

Mansour, E. H., and A. H. Khalil. 2000. Evaluation of antioxidant activity of some plant extracts and their application to ground beef patties. *Food Chem* 69:135–41.

Maybey, R., and M. Mclintyre. 1988. *The New Age Herbalist: How to Use Herbs for Healing, Nutrition, Body Cafre and Relaxation*. New York: Collier, p. 80.

Menzies, I. S., M. J. Zuckerman, W. S. Nukajam et al. 1999. Geography of intestinal permeability and absorption. *Gut* 44 (4):483–9.

Milner, J. A. 1996. Garlic: its anticarcinogenic and antitumorigenic properties. *Nutr Rev* 54 (11):S82–S83.

Mishra, A. K., and N. K. Dubey. 1990. Fungitoxic properties of Prunus persica oil. *Hindustan Antibiot Bull* 32:91–3.

Nagourney, R. 1998. Garlic: medicinal food or nutritious medicine? *J Med Food* 1:13–28.

Oda, Y. 1995. Inhibitory effect of curcumin on SOS functions induced by UV irradiation. *Mutat Res* 348 (2):67–73.

Ody, P. 1993. *Allium sativum, Clovers: The Complete Medicinal Herbal*, p. 33.

Ody, P. 1993. *Allium sativum, Garlic and Cloves: The Complete Medicinal Herbal*, p. 33.

Ody, P. 1993. *Allium sativum, Garlic: The Complete Medicinal Herbal*, p. 33.

Pari, L., and A. Uma. 2003. Protective effect of Sesbania grandiflora against erythromycin estolate-induced hepatotoxicity. *Pharmacologie* 58 (5):439–43.

Piccaglia, R., M. Marotti, E. Giovanelli et al. 1993. Antibacterial and antioxidant properties of mediterranean aromatic plants. *Ind Crop Prod* 2 (1):47–50.

Qin, J., R. Li, J. Raes et al. 2010. A human gut microbial gene catalogue established by metagenomic sequencing. *Nature* 464: 59–65.

Ramezani, A., and D. S. Raj. 2014. The gut microbiome, kidney disease, and targeted interventions. *J Am Soc Nephrol* 25 (4):657–70.

Roman-Ramos, P., J. L. Flores-Saenz, and F. J. Alarcon-Aguilar. 1995. Anti-hyperglycemic effect of some edible plants. *J Ethanopharmacol* 48:25–32.

Roth, G. N., A. Chandra, and M.G. Nair. 1998. Novel bioactivities of Curcuma longa constituents. *J Nat Prod* 61:542–5.

Ruberto, G., M. T. Baratta, S. G. Deans et al. 2000. Antioxidant and antimicrobial activity of Foeniculum vulgare and Crithmum maritimum essential oils. *Planta Med* 66:687–93.

Sahu, B. R., and A. Chakrabarty. 1994. Screening of antibacterial activity of various extractives of seeds of Cassia tora and alternanthera sessilis. *Asian J Chem* 6 (3):687–9.

Sartor, R. B. 2008. Microbial influences in inflammatory bowel diseases. *Gastroenterology* 134 (2):577–94.

Sathyabama, S., N. Khan, and J. N. Agrewala. 2014. Friendly pathogens: prevent or provoke autoimmunity. *Crit Rev Microbiol* 40 (3):273–80.

Shams-Ghahfarokhi, M., M.-R. Shokoohamiri, N. Amirrajab et al. 2006. In vitro antifungal activities of Allium cepa, Allium sativum and ketoconazole against some pathogenic yeasts and dermatophytes. *Fitoterapia* 77 (4):321–3.

Sharma, A., S. Gautam, and S. S. Jadhav. 2000. Spice extracts as dose-modifying factors in radiation inactivation of bacteria. *J Agric Food Chem* 48 (4):1340–4.

Singh, G., P. Marimuthu, H. S. Murali et al. 2005. Antioxidative and antibacterial potentials of essential oils and extracts isolated from various spice materials. *J Food Saf* 25 (2):130–45.

Solis, C. 1969. Antibacterial and antibiotic properties of the leguminosae. *Acta Manila Ser A* 4:53–109.

Spiliotis, V., S. Lulas, V. Gergis et al. 1998. Comparison of antimicrobial activity of seeds of different Moringa oleifera varieties. *Pharm Pharmacol Lett* 8 (1):39–40.

Sugita, R., E. Hata, A. Miki et al. 2012. Gut Colonization by Candida albicans inhibits the induction of humoral immune tolerance to dietary antigen in BALB/c mice. *Biosci Microbiota Food Health* 31 (4):77–84.

Suksmrarnh, S., N. Suwannapoch, W. M. Phakhodee et al. 2003. Antimycobacterial activity of prenylated xanthones from the fruits of Garcinia mangostana. *Chem Pharm Bull* 51 (7):857–9.

Tkachenko, K. G., N. V. Kazarinova, L. M. Muzyehenko et al. 1999. Antibiotic properties of essential oils of some plant species. *Nauka* 35:11–23.

Tomasi, F., G. Cammpadelli-Fiume, L. Barbieri et al. 1982. Effect of ribosome-inactivating proteins on virus-infected cells. Inhibition of virus multiplication and of protein synthesis. *Arch Virol* 71:323–32.

Vaishampayan, P. A., E. Rabbow, G. Horneck, and K. J. Venkateswaran. 2012. Survival of Bacillus pumilus spores for a prolonged period of time in real space conditions. *Astrobiology* 12 (5):487–97.

Vijayakumar, R. S., D. Surya, and N. Nalini. 2004. Antioxidant efficacy of black pepper (Dorman and Deans) and piperine in rats with high fat diet induced oxidative stress. *Redox Report* 9 (2):105–10.

Yamamoto, T., I. A. Watkinson, L. Kim et al. 1988. Nucleotide sequence of the gene coding for a 130-kDa mosquitocidal protein of Bacillus thuringiensis israelensis. *Gene* 66 (1):107–20.

Yamashiki, M., A. Nishimura, X. Huang et al. 1999. Effects of the Japanese herbal medicine "Sho-saiko-to" (TJ-9) on interleukin-12 production in patients with HCV-positive liver cirrhosis. *Dev. Immunol* 7 (1):17–22.

Yesilada, E., I. Gurbuz, and H. Shibata. 1999. Screening of Turkish anti-ulcerogenic folk remedies for anti-Helicobacter pylori activity. *J Ethanopharmacol* 66:289–93.

Yoo, D. H., I. S. Kim, T. K. Van Le, I. H. Jung, H. H. Yoo, and D. H. Kim. 2014. Gut microbiota-mediated drug interactions between lovastatin and antibiotics. *Drug Metab Dispos* 42(9):1508–13.

Youdim, K. A., and S. G. Deans. 1999. Beneficial effects of thyme oil on age-related changes in the phospholipid C20 and C22 polyunsaturated fatty acid composition of various rat tissues. *Biochim Biophys Acta* 1438 (1):140–6.

Zheng, Y. T., K. L. Ben, and S. W. Jin. 1999. Alpha-momorcharin inhibits HIV-1 replication in acutely but not chronically infected T-lymphocytes. *Acta Pharmacol Sin* 20 (3):239–43.

6 Interactions of Natural Products with Genomes
Future Challenges

Siva G. Somasundaram

CONTENTS

6.1   INTRODUCTION

There are 214,000 natural products that have been reported in the Dictionary of Natural Products (DNP) database (http://dnp.chemnetbase.com). It is necessary to investigate the biological activities of these natural products with reference to genomic interactions. So far, not much has been done. In this chapter, an attempt has been made to elucidate the protocol of planning and performing the genomic interactions from bacteria to human model systems.

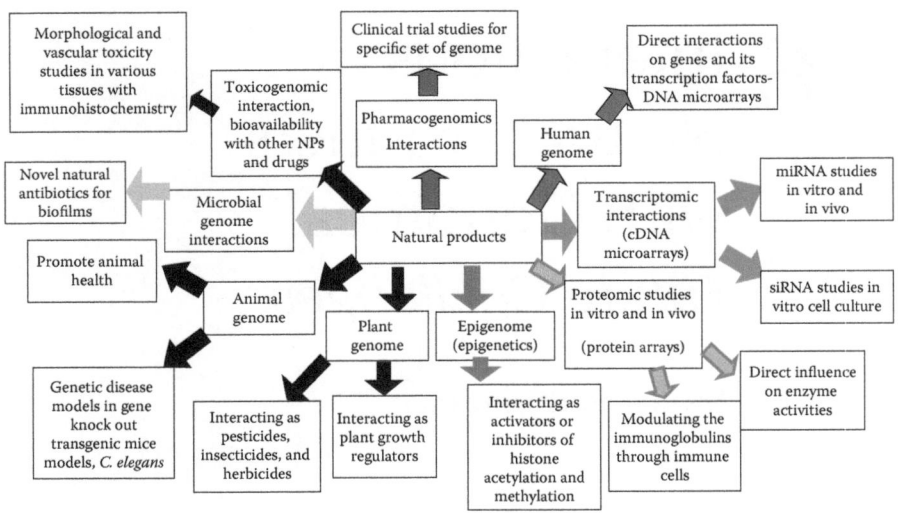

FIGURE 6.1 **(See color insert.)** A model flowchart for natural products interacting with various genomes.

Figure 6.1 explains interactions of natural products with different levels of genomic studies. In this book, we have attempted to address a part of the interaction pathways of "natural product interactome." Chapter 2 addressed most of the studies that are available for specific genes and selected natural products. We have not studied all the genes that are present in each chromosome but instead have designed an analytical approach from the available literature for the interaction of human genome. Chapter 3 addressed most of the transcriptomic profiles of orange juice and hesperidin, a natural product. The second part of this chapter addressed the specific functional genome, that is, metastasis genome interaction with limonin, a citrus natural product in animal model studies. In Chapter 4, we discussed an interaction pathway where the advantages and disadvantages of combinations of traditional therapy or interactions of different natural products on both breast cancer cells and prostate cancer cells with specific reference to the signaling kinases activities are present. In addition, the immunohistopathological studies also provide further confirmation of the effectiveness of the combinatorial therapy in vascular tissue differentiations.

In Chapter 5, we addressed the protective activity of interactions of natural products against radiotherapy-induced oxidative stress gene activity. Chapter 6 addressed 139 natural products on various microorganisms' growths and antioxidative activity. In this chapter, we tend to simplify the genomic interactions by the structural similarity of the compounds and the genes. An attempt has been made to collect and present the relevant literatures for interactions of natural products with epigenetic mechanisms on cell lines and animal models through modulating microRNA expressions. Also, we explore the genetic interactions of natural products derived from marine organisms, animals, and humans. Finally, we present the effect of natural products on cytochrome P450 genes for the interaction of prescribed drug and natural products.

6.2 STRUCTURAL BASIS OF INTERACTIONS OF NATURAL PRODUCTS WITH GENES

There are several reports that address the effects of natural products on reducing experimentally induced inflammation using in vitro as well as in vivo animal models. None of them address the genomic profile of inflammation with specific natural products. The combination of flavanoidal glycosides such as apigenin, scutallerin, and pectinolinergenin inhibits experimentally induced arthritis (Siva and Blessy, 2013). It has been mentioned that inhibiting the COX1 and COX2 enzymes along with the lipoxygenases enzymes and at the same time enhancing the superoxide dismutase and glutathione reductase may offer a long-lasting effect for experimentally induced arthritis models without any side effects when compared to steroidal therapy (Figure 6.2).

The mechanism of action depends on the *ortho*-dihydroxy groups that directly influence the respective gene activity of enzyme proteins. We found from the PubChem database that there are 31 groups of flavonoids. When we analyzed the structures, we found that there are six groups of flavonoids, quercetin, catechin, apigenin, luteolin, rutin, and genistein, in *ortho*-dihydroxy groups. The total number of 32,683 similar compounds are available from all these six groups (Figure 6.3a through f). We thereby suggest that these 32,683 compounds may share a common mechanism of interaction with the same genes that have been reported to be activated by apigenin. Hence, we can extend the similar mechanism so that *ortho*-dihydroxy flavonoids and their derivatives may follow the same pattern of activities in our analyses in Chapter 2. In the same way, we need to address the chemical structural basis of interactions for 214,000 natural products from the DNP with specific genes in the future. This study will become the next milestone in the research of the interaction of natural products and genome.

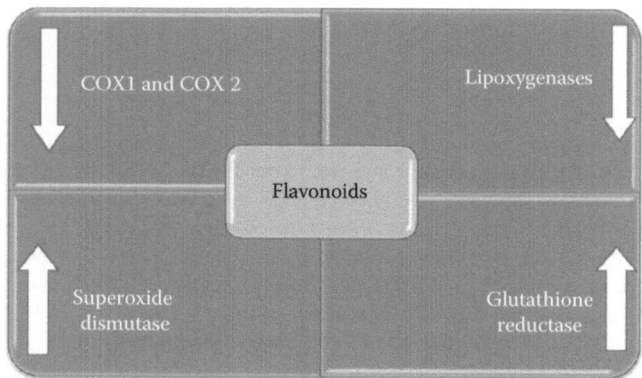

FIGURE 6.2 **(See color insert.)** A chart explains the role of flavonoids on both antiinflammatory and antioxidative mechanisms by increasing gene activities for antioxidant enzymes and decreasing gene activity of prostaglandin pathways. (Data from Somasundaram, S., and Oommen, B., *Antioxidant Flavonoids for Arthritis Treatment: Human and Animal Models*, edited by R. R. Watson and V. R. Preedy, 1st Edition, Elsevier, London, 2013.)

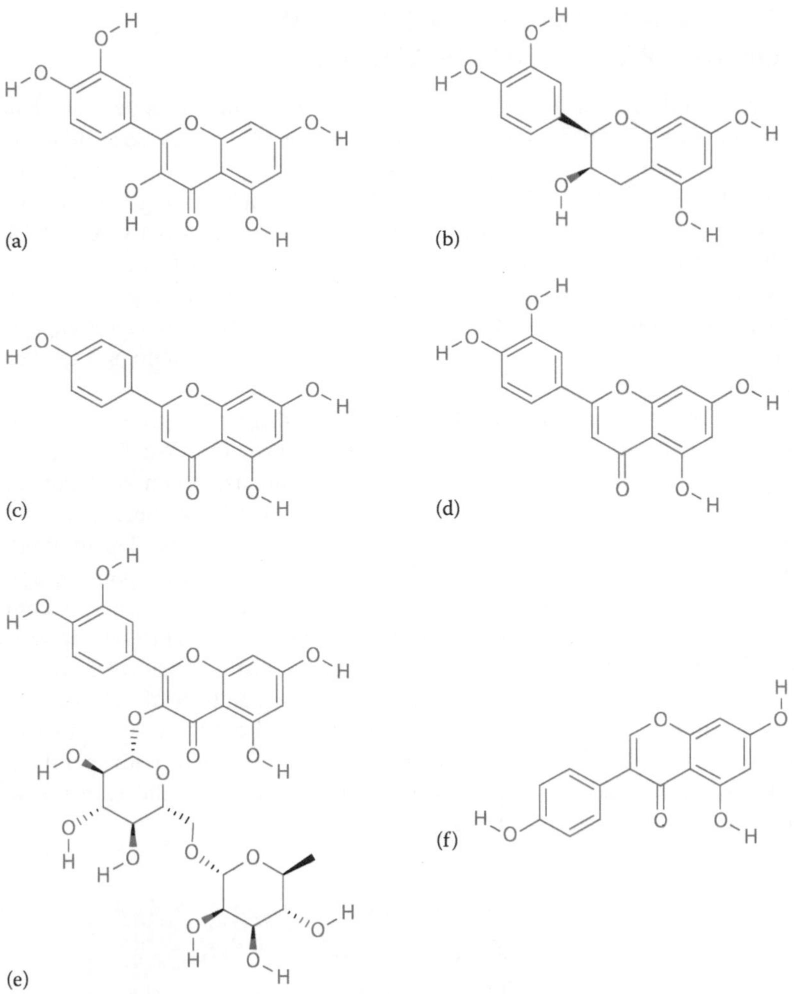

FIGURE 6.3 (a) Structure of quercetin. IUPAC: 2-(3,4-dihydroxyphenyl)-3,5,7-trihydroxychromen-4-one, number of similar structural compounds available are 3818. (b) Structure of catechin. IUPAC: 2R,3R-2-(3,4-dihydroxyphenyl)-3,4-dihydro-2H-chromene-3,5, number of similar structural compounds available are 2210. (c) Structure of apigenin. IUPAC: 5,7-dihydroxy-2-(4-hydroxyphenyl)-4H-chromen-4-on, number of similar structural compounds available are 5984. (d) Structure of luteolin. IUPAC: 5,7,3′,4′-tetrahydroxy-flavone, number of similar structural compounds available are 4432. (e) Structure of rutin. IUPAC: 2-(3,4-dihydroxyphenyl)-5,7-dihydroxy-3-[(2S,3R,4S,5S,6R)-3], number of similar structural compounds are available 10,775. (f) Structure of genistein. IUPAC: 5,7-dihydroxy-3-(4-hydroxyphenyl)chromen-4-one, number of similar structural compounds available are 5464.

6.3 NATURAL PRODUCTS INTERACTING WITH GENES THROUGH MICRORNAS

Post transcriptional modifications are done by short non-coding microRNAs (miRs), and these miRs regulate the expression of several target genes in the genome (Bartel, 2004) by binding to 3′UTR regions and inhibiting the translational process. It is very important to explore the effect of natural products on the posttranscriptional modifications of genes through miRs. There are only a few natural products that have been studied in detail. We expect in the future that more studies will be available for the interactions of various natural products with miRs as an epigenetic mechanism to address different important gene expressions.

6.3.1 CURCUMIN INTERACTING MIRS

The natural product curcumin inhibits the proliferative effects of bisphenol A on MCF-7 cells through the modulation of a miR-19/PTEN/AKT/p53 pathway. In this study, curcumin downregulates the stimulated miR-19a and miR-19b in MCF-7 cells (Li et al., 2014). In another study, curcumin upregulates miR-181b and downregulates CXCL1 and -2 in cells isolated from primary human breast cancers (Kronski et al., 2014). In addition, in hepatic stellate cells, curcumin increases the miR-29b, which hypomethylates phosphatase and tensin homologs (PTEN) that specifically act toward DNA methylase downregulation (Zheng et al., 2014). Another miR, miR-21, is increased in cancer and named as *oncomir*. Curcumin analog inhibits this miR-21 and does not affect the expression of other miR-100, miR-181a, and miR-200a in prostate cancers (Yang et al., 2013). A genome wide study indicates that curcumin increased the expression of antioxidant genes and reduced angiotensin II type 1 receptor, nuclear factor-kappa B, and vascular endothelial growth factor expression at the messenger RNA and protein levels. In this study, curcumin downregulates 20 miRs and upregulates 9 genes in ARPE-19 cells exposed to oxidative stress (Howell et al., 2013). Also, curcumin modulates miRNA expression in human pancreatic cells. It upregulates miRNA-22 and downregulates miRNA-199a and exerts its anticancer activity (Sun et al., 2008).

6.3.2 GENISTEIN INTERACTING MIRS

Genistein downregulates the gene expression of miR-151 in PC3 and DU145 prostate cancer cells (Chiyomaru et al., 2012). In another study, the miR-574-3p level is decreased in prostate cancer cell lines and prostate cancer tissues when compared to adjacent normal tissues and normal prostate cells (Chiyomaru et al., 2013). This low level of miR-574-3p is correlated with the advanced tumor stage, and miR-574-3p is a tumor suppressor gene. Genistein increases miR-574-3p levels in prostate cancer cell lines and suppresses cancer cell proliferation (Chiyomaru et al., 2013). Another miR, miR-34a, is upregulated by genistein and inhibits its target oncogenic HOTAIR in prostate cancer cell lines, PC3, and DU145 (Chiyomaru et al., 2013). In pancreatic cancer cells, genistein upregulates miR-34a that targets the downregulation of Notch-1 and leads to the inhibition of cell growth and induction of apoptosis (Xia et al., 2012). In addition, genistein inhibits miR-223 expression and upregulates its target gene Fbw7 in pancreatic cancer cell lines and induces apoptosis (Ma et al., 2013). Also, genistein inhibits the expression of miR-27a in pancreatic cancer cells and induces apoptosis (Xia et al., 2014). In renal cell carcinoma, genistein inhibits the expression of miR-23b-3p in RCC cell lines. This miR-23b-3p, an oncomir, is elevated during the carcinogenesis and inhibits the tumor suppressor gene PTEN. As genistein inhibits the expression of miR-23b-3p, it enhances the activation of the tumor suppressor gene PTEN, thereby inhibiting cell proliferation (Zaman et al., 2012). The expression of miR-1260b was significantly higher in renal cancer tissues compared with normal tissues, and it inhibits its targeted genes expressions such as sFRP1, Dkk2, and Smad4. Genistein inhibits miR-1260b and enhances the expression of its target genes (Hirata et al., 2013). The expression of miR-27a is elevated in human ovarian cancer when compared to benign ovarian tissues. Genistein downregulates this miR-27a expression and stimulates the target gene expression Sprouty2 and induces apoptosis in ovarian cancer (Xu et al., 2013).

6.3.3 RESVERATROL INTERACTING MIRS

In lung cancer, the natural product resveratrol plays an indirect role that depends on a specific miR and then modifies RECK gene activity. That is, resveratrol acts only on miR-200c-positive cells and not on miR-200c-negative cells (Bai, Dong, and Pei, 2014). In colon cancer, resveratrol increased the intracellular expression level of miR-34a and downregulated the target gene E2F3 and its downstream Sirt1 cascade to induce apoptosis (Kumazaki et al., 2013).

The combinatorial effects of natural products have an impact on miR levels. Resveratrol and quercetin in the mixture of a 1:1 ratio decreased miR-27a and induced zinc-finger protein ZBTB10, a tumor repressor protein in HT29 colon cancer cell line that is beneficial in apoptosis through activating caspase-3 cascade and increased PARP cleavage (Del Follo-Martinez et al., 2013). In the central nervous system and aging disorders, resveratrol plays a significant role in regulating miR levels. A recent study indicates that resveratrol reduces the expression of miR-134 and upregulates cyclic AMP response element-binding protein (CREB) levels to initiate the synthesis of brain-derived neurotrophic factor (BDNF) that is required for neurotransmitter synthesis involved in learning, cognition, and self-esteem (Zhao et al., 2013).

6.3.4 QUERCETIN INTERACTING MIRS

Quercetin and its major metabolites interact with proinflammatory genes through miRs. Proinflammatory miR-155 is downregulated by quercetin and isorhamnetin, but not by quercetin-3-glucuronide in murine RAW264.7 macrophages stimulated with lipopolysaccharides, indicating that the role of the parent natural product is different from the metabolites in controlling the gene expression through miRs (Boesch-Saadatmandi et al., 2011). In another study, rhamnetin is involved in sensitizing radiotherapy resistance lung cancer cells for apoptosis. Rhamnetin enhances the tumor suppressive miR-34a level and thereby increases radiotherapeutic efficacy by inhibiting radiation-induced Notch-1 signaling associated with radioresistance in non-small cell lung cancer (Kang et al., 2013).

6.3.5 Caffeic Acid Phenyl Ester Interacting miRs

Alzheimer's disease (AD) is the progressive accumulation of amyloid beta peptides in the brain. The "triggering receptor expressed in myeloid cells 2" (TERM2) is essential in the sensing, recognition, phagocytosis, and clearance of noxious cellular debris from brain cells, including neurotoxic Aβ42 peptides. A mutation in this gene causes amyloid deposition. In an experimental condition, aluminum-sulfate, when incubated with microglial cells, induces the upregulation of an NF-κB-sensitive miR-34a that is known to target the TREM2 miR-3′-untranslated region and downregulate TREM2 expression. However, caffeic acid phenyl ester downregulates miR-34 and thereby enhances TERM2 expression (Alexandrov et al., 2013) and inhibits amyloid deposition.

6.3.6 Capsaicin Interacting miRs

There are only two reports available for capsaicin activity on miR interactions. In K562 leukemic cell proliferation, capsaicin plays a significant role to exhibit its anti-cancer property through interaction with miR-520a-5p/STAT3 and induces apoptosis. In this study, capsaicin downregulates miR-520a-5p expression and decreases STAT3 protein levels (Kaymaz et al., 2014). Capsaicin displays a differential miR expression activity in the nervous system of mice pain models. Capsaicin increases the expression of miR-1 and miR-16 in dorsal root ganglion cells and decreases the expression of miR-206 in spinal dorsal horns (Kusuda et al., 2011).

6.3.7 Apigenin Interacting miRs

There are only two reports available for apigenin activity on miR interactions. In the transgenic mice model, the administration of apigenin suppressed the matured miR-103 expression levels and improved glucose tolerance (Ohno et al., 2013). In neuroblastoma cells, there is a decrease in expression of tumor suppressor miR-138 that targets telomerase gene activity. Apigenin enhances miR-138 and inhibits cell growth (Chakrabarti, Banik, and Ray, 2013).

6.3.8 Lycopene Interacting miRs

Lycopene plays an important role in stearic acid–induced liver disorders. MiR-21 expression decreases during the experimentally induced liver disorder in mice and Hepa cells treated with stearic acid. Lycopene upregulates mirR-21 and decreases its target gene fatty acid-binding protein 7 activities, in both transcriptional and translational levels, which help hepatic lipid metabolism in nonalcoholic fatty liver disease (Ahn et al., 2012).

6.3.9   BETULINIC ACID INTERACTING MIRS

Betulinic acid, a derivative of betulin that is a triterpene from birch trees (Alakurtti et al., 2006), induces apoptosis and inhibits cell growth in Erb-expressing BT-474 breast cancer cells. The mechanism of action includes the betulinic acid downregulation of miR-27a inducing of the specificity protein repressor transcription factor ZBTB10 and downregulation of Sp1, Sp3, and Sp4 gene expression, which thereby decreases the ErbB2 expression (Liu et al., 2012). Further, the major mechanism of Sp downregulation by betulinic acid is through the disruption of miR-27a: ZBTB10, which is cannabinoid receptor dependent, and betulinic acid directly bind to both CB1 and CB2 receptors.

6.3.10   CANNABINOIDS INTERACTING MIRS

Tetrahydrocannabinol, an active metabolite of cannabinoid, induces miR-690 and targets the gene for transcription factor CCAAT/enhancer-binding protein α (C/EBPα), which is involved in the immunosuppression of myeloid-derived suppressor cells (Hegde et al., 2013). There are two receptors for cannabinoid CB1, which is present mostly in the central nervous system, while CB2 is present mostly in immune cells. Other ligands, such as GPR55, are present in both CNS and immune cells and also in the intestine, the bone marrow, and the spleen. Cancer cells express CB1, CB2, and other GPR ligands receptors (Matsuda et al., 1990). These receptors can bind not only with cannabinoids and endocannabinoids but also with natural products such as betulinic acid and inhibit cell growth. Endocannabinaoids are synthesized in the body and responsible for analgesic activity. Popular prescribed analgesic medications exhibit their activities through elevating endocannabinoids. Endocannabinoids bind readily with CB1 and CB2 receptors and degrades. However, metabolites of paracetamol and N-arachidonoylaminophenol are a weak agonist to CB1 and CB2 and inhibit the reuptake of endocannabinoides and increase endocannabinoide levels in the tissues and increase the analgesic efficacy of paracetamol (Bertolini et al., 2006; Sinning et al., 2008). Another natural product, cannabinoid's metabolite cannobitol, also inhibits CB1 and CB1 receptors resulting in an increase in the levels of endocannabinoid in the brain and exerting its antipsychotic effects in the treatment of schizophrenia (Leweke et al., 2012).

6.3.11   EPIGALLOCATECHIN GALLATE INTERACTING MIRS

Green tea, a natural product, and epigallocatechin gallate (EGCG) upregulate miR-16 and target the Bcl-2 gene, which is an anti-apoptotic protein. Thus, EGCG treatment enhances the suppression of the anti-apoptotic process and induces apoptosis in HepG2 cells (Tsang and Kwok, 2010). EGCG enhances the expression of miR-7-1 for the induction of apoptosis in human malignant neuroblastoma cells (Chakrabarti et al., 2013). Another study result indicates that 6 miRs are upregulated and 26 miRs are downregulated with an EGCG treatment in UV-exposed normal human dermal fibroblasts. The upregulations of miRs are in miR-1246, -145, -4299, -548c-3p, -636, and -933. The downregulations of miRs are in miR-BART12, -BART1-1, -1202,

-1207, -1207-5p, -1225-5p, -1227, -1271, -133(a), -134, -181d, -212, -3141, -362-3p, -3667-5p, -3679-3p, -3907, -423-3p, -4270, -455-5p, -494, -513a-5p, -660, -718, and -K12-10b (An, 2013). Also, EGCG upregulates the tumor-suppressor miRs let-7a-1 and let-7g in lung cancer cell lines and induces apoptosis (Zhong et al., 2012). EGCG upregulates the expression of miR-210 and leads to the decrease in cell proliferation rate in lung cancer cells through stabilization of hypoxia-response element protein (Wang, Bian, and Yang, 2011). Green tea extract, polyphenon-60, downregulates miR-21 and miR-27 in breast cancer cells, the target gene of a tumor suppressor gene, tropomyosin-1, and increases anticancer potential (Fix et al., 2010).

6.4 NATURAL PRODUCTS FROM MARINE ORGANISMS INTERACTING WITH GENES

About 70% of the Earth is surrounded by ocean. Marine organisms such as bacteria, algae, soft corals, sponges, mollusks, phytoplanktons, tunicates (ascidians), and echinoderms produce compounds that are useful for various pharmaceutics. These compounds are very useful as antibacterial, analgesic, anti-inflammatory, antimalarial, anticancer, antiparasitic, and antiviral agents (Shushizadeh, 2014). Natural products play a significant role in drug discovery. Almost 42% of anticancer drugs are from natural products or their derivatives (Newman and Cragg, 2007). Also, 22,000 biologically active compounds have been obtained from microbial resources. Actinobacteria alone contributes 45% of microbial natural products (Bérdy, 2005). The latest development in pursuing this novel is that natural product compounds are emerging from marine organisms. The secondary metabolites formed from marine invertebrates such as sponges, tunicates, and nudibranchs provide promising natural products for drug therapy.

At present, there are three natural products from marine sources that have been approved for regular therapy (Molinski and Morinaka, 2012). They are (1) ziconotide, (Prialt®, 1), a "cysteine knot" peptide isolated from the cone snail *Conus magus* used for neuropathic pain; (2) Yondelis® (Trabectidin, ET-743, 2), a complex tetrahydroisoquinoline alkaloid isolated from the Caribbean tunicate *Ecteinascidia turbinate* used for soft tissue sarcoma; and (3) halichondrin B, a polyether toxin isolated from marine sponge *Halichondria okadai* used as a scaffold for synthetic refinements to the newly licensed drug.

The molecular biology of the interaction of ziconotide on genes is quite different from the traditional morphine therapy as a pain killer during chronic pain. In this example, the presynaptic voltage-gated calcium $Ca_v2.2$ channels play a significant role in spinal sensitization during peripheral injury. It is reported (Molinski and Morinaka, 2012) that a pair of genes, *cancna1b* gene e37a and e37b, is involved to maintain the efficacy of the calcium channel $Ca_v2.2$ by morphine. In contrast, the natural product isolated from a marine organism, ziconotide or gabapentin, is not participating in alternate pre-mRNA splicing for the *Cacna1b* exon e37a or e37b.

Another natural product from a marine source is trabectedin, which interferes with the aberrant transcription mechanism and is more active in nucleotide excision repair and homologous recombination repair-deficient cells (García et al., 2013).

When treated to these cells, this natural product interacts with the gene CUL4A and induces its expression and thereby decreases the BRCA1/ERCC5, BRCA1/CUL4A, and XRCC3/CUL4A expression ratios. Thus, it is a new developing drug for solid breast tumors expressing BRCA genes (García et al., 2013).

The next study focused on the clinical trial for anticancer drugs from marine natural products. They products are from eubacteria through eukaryotes such as fungi and protists but none in archaea for anticancer property (Newman and Cragg, 2014). One of the clinical trial drugs for cancer is cytarabine, which is a cytosine arabinoside, and is used alone or in combination with other antineoplastic drugs for leukemia. This is an antimetabolite drug and incorporates into human DNA and consequently inhibits leukemic cell by interacting with DNA and RNA synthesis. Unlike the regular conventional chemotherapeutic drugs, cytarabine induces autophage only in leukemic cell lines and primary leukemic cells and not in healthy leukocytes (Newman and Cragg, 2014). This is an advantage for using natural products over conventional chemotherapy because conventional chemotherapy kills both cancer and healthy cells and leads to toxic side effects.

The genomic interactions of two potent anticancer tetrahydroisoquinoline alkaloids, namely Trabectedin and Zalypsis, can directly form a covalent bond with the amino group of a guanine in selected gene triplets of DNA duplexes and lead to double-strand breaks (Feuerhahn et al., 2011). The location of the cleavage by the XPF/ERCC1 nuclease is in the opposite strand bonding with the drug. The mode of action includes that the drug binds to one strand and relaxes the opposite strand to enable the binding of nuclease to the opposite strand. Also, the two marine natural product drugs can prevent the binding of transcription factors like Sp1 to DNA and inhibit the elongation of RNA polymerase II at the same nucleotide.

6.5 INTERACTIONS OF NATURAL PRODUCTS FROM ANIMALS AND HUMANS WITH GENES

Certain traditional medicines from ayurvedic (Dev, 1999) as well as Chinese systems of medicine reported that several natural products originating from animals have healing properties for various illnesses (Ji, Li, and Zhang, 2009). The genomic interactions of these natural products are yet to be investigated. There are six animal natural products that are reported to have a beneficial effect on humans (Mahomoodally and Sreekeesoon, 2014). They are *Bos taurus* (cow ghee), *Apis mellifera* (honeybee), *Crassostrea* spp. (oyster), *Donax trunculus* (bivalve), *Helix aspersa* (Shankar et al., 2011), and *Coturnix japonica* (quail egg). They are used for anti-inflammatory and antiasthmatic activities on Mauritius island.

In addition, there are 19 natural products that have been claimed for useful therapeutic agents from the animal kingdom. These animals include bivalves, camel, crab, goat, hard shelled turtle, honeybees, human, peacock, pigeons, sambhar, sheep, and snails (Mahawar and Jaroli, 2007). From Latin America, there are 584 zootherapies that are used (Figure 6.4) (Alves and Alves, 2011). The genomic interactions of these natural products are yet to be investigated.

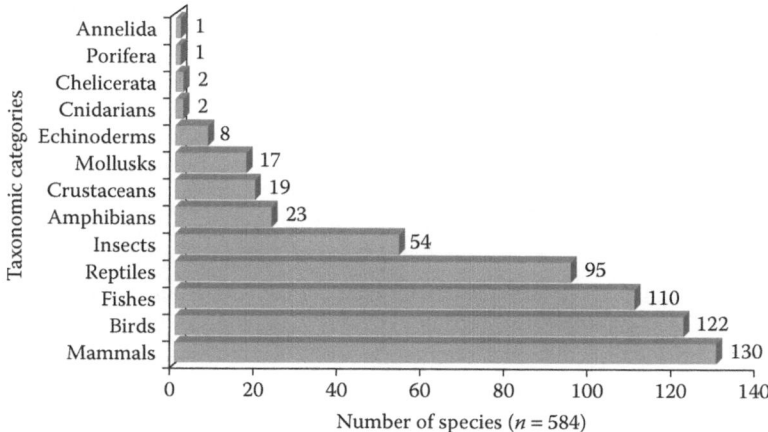

FIGURE 6.4 **(See color insert.)** Number of animal species used as remedies per taxonomic category in Latin America. (Data from Alves, R.R., and Alves, H.N., *J. Ethnobiol. Ethnomed.*, 7, 9, 2011. Published online March 7, 2011. doi: 10.1186/1746-4269-7-9. Copyright 2011 Alves and Alves; licensee BioMed Central Ltd. This is an open-access article distributed under the terms of the Creative Commons Attribution License, http://creativecommons.org/licenses/by/2.0, which permits unrestricted use, distribution, and reproduction in any medium, provided the original work is properly cited.)

Another report stated about leech therapy. There are several species used for medicinal purposes. They are *Hirudo medicinalis, H. orientalis, H. troctina, H. verbena, Hirudinaria manillensis,* and *Macrobdella decora.* Leech therapy has been used for the conditions such as osteoarthritis, venous congestion, and surgical reconstructions (Heckmann et al., 2005; Michalsen et al., 2003; Whitaker et al., 2012). In addition, leech therapy is used in the cases of nevus pigmentation management (Rastogi and Chaudhari, 2014). Again, the genomic interactions of the natural products from leeches are yet to be studied.

The natural product from camel milk has a profound effect on insulin-like activity (Korish, 2014). The study indicates that camel milk acts on experimentally induced diabetes animal models to lower the fasting glucose level, hypolipidemia, decreased HOMA-IR, recovery of insulin secretion, weight gain, and no mortality during the study. Camel milk also inhibits the diabetes-induced elevation in pancreatin hormones TNF-α and TGF-β1 levels. In another study, camel milk inhibits high cholesterol diet–induced nonalcoholic fatty liver disease through its antioxidative property of natural products (Korish and Arafah, 2013). Camel milk is enriched in minerals, vitamins, insulin, and insulin-like protein. These ingredients are directly involved in lowering the liver fat deposition, decreasing the infiltration of inflammatory cells, preserving liver function, increasing the reduced glutathione level by stimulating glutathione reductase gene and stimulating catalase activity, and decreasing the malondialdehyde levels in experimental animals (Korish and Arafah, 2013). Much work on other genes is yet to be investigated.

Human breast milk has the natural product that stimulates immune activity. Colostrum is known as *liquid God* for newborn babies (Pletsch et al., 2013). In addition, human milk oligosaccharides as well as bovine milk oligosaccharides have unique interactions with genes of intestinal epithelial cells to boost the immune system (Lane et al., 2013). The oligosaccharides interact with the battery of genes involved in the expression of cytokines IL-17C and platelet factor 4, chemokines (chemokine [C-X-C motif] ligand 1 [CXCL1], chemokine [C-X-C motif] ligand 3, chemokine ligand 20, chemokine [C-X-C motif] ligand 2 [CXCL2], chemokine [C-X-C motif] ligand 6, chemokine ligand 5, chemokine ligand 1 [CX3CL1], and CX3CL2) and cell surface receptors, intercellular adhesion molecule-1 (ICAM-1), intercellular adhesion molecule-2 (ICAM-2), and IL-10 receptor α. This section of animal-originated natural products has a unique action on human genome. More research is needed for the interactions of animal natural products with the human genome.

6.6   INTERACTIONS OF NATURAL PRODUCTS WITH CYTOCHROME P450 SYSTEMS

Quassin and neoquassin from *Picrasma excels* inhibits CYP1A1, once the isoenzymes of cytochome P450, which is an activator of carcinogen (Shields et al., 2009). CYP1A2 metabolizes important drugs such as theophylline, caffeine, imipramine, and propranolol (Zhai et al., 1998). Also, there is some difference in the inhibitory activity between natural products apigenin and α-naphthoflavone on CYP1A1 and CYP1A2 (Pastrakuljic et al., 1997). The flavonoid that has the 3-, 5-, and 7-trihydroxylation inhibits CYP1A2 activity to a greater extent than hydroxylation at position 3 or 5 alone. However, the binding environment of the CYP1A1 active site has a preference for the 7-hydroxyl substituent.

Butanamide from *Amyris plumieri* inhibits CYP1A1 and exhibits anticancer activity in breast cancer cells (Badal et al., 2011). Curcumin inhibits CYP3A4, a drug metabolizing enzyme for tamoxifen in the liver and the small intestine. Curcumin also inhibits P-glycoprotein drug efflux transporter in the small intestine and increases the bioavailability of tamoxifen for its anticancer activity (Cho et al., 2012). In contrast, CYP2D6 enzyme treated curcumin with copper (II) induces a carcinogen, 8-oxo-7 8-dihydro-2(′)-deoxyguanosine through O-demethyl curcumin and elevates the carcinogenic potency (Sakano et al., 2002). Hence, much work need to focus on natural products and their choice of cytochrome P system targets. Limonin, a citrus natural product, does not inhibit human CYP1A2, CYP2C8, CYP2C9, CYP2C19, CYP2D6, and P-gp. However, it does inhibit CYP3A4 isoenzyme. It has important implications with regard to natural product–drug interactions when a drug activity depends on CYP3A4 isoenzyme (Han et al., 2011). Oramucosal cannabinoids and tetrahydrocannabinol use CYP3A4 isoenzyme for their metabolism (Stout and Cimino, 2014). If this enzyme is inhibited by limonin, then cannabinoids and tetrahydrocannabinol may not be metabolized. On the other hand, the smoked cannabis induces CYP1A2 (Stout and Cimino, 2014), but apigenin inhibits CYP1A2 gene activity. This differential interaction indicates that the route of administration of cannabinoids may have an impact on dietary-originated natural products on the cytochrome P450 system. The human liver microsomes cytochrome P450 isoenzyme study reveals the cannbinoids oxidations. This study again shows that CYP3A4 and CYP2C19 may be major isoforms responsible for 6α-, 6β-, 7-, and/or 4″-hydroxylations of cannabinoids (Jiang et al., 2011). Therefore, caution should be exercised when we use cannabinoids as therapeutic intervention while choosing diets with specific natural products to minimize the side effects and maximize the therapeutic activities.

6.7 CONCLUSION

Figure 6.1 summarizes the present and future perspectives of the interactions of natural products with the genome. We have presented plant, microbial, marine, animal, and human natural products and their interactions on various genes. There are thousands of natural products that have yet to be studied with reference to the human genome and other genomes. Natural products and prescribed drug interactions on the drug metabolizing genetic system are very important in terms of diet and therapeutic efficacy. The latest explosion of miRs research is now in its infant stage as far as interactions of natural products are concerned. Hence, future research should focus on interactions of natural products toward the transcriptomics approach through miRs. There are more challenges in terms of genome-wide interactions of natural products. All the data regarding specific gene interactions with multiple natural products may be documented in a database for retrieval to advance the knowledge. Perhaps, in the next edition we will provide additional information and more comprehensive data for natural products–genome interactions.

REFERENCES

Ahn, J., H. Lee, C. H. Jung, and T. Ha. 2012. Lycopene inhibits hepatic steatosis via microRNA-21-induced downregulation of fatty acid-binding protein 7 in mice fed a high-fat diet. *Mol Nutr Food Res* 56 (11):1665–74.

An, I. S., S. An, S. Park, S. N. Lee, and S. Bae. 2013. Involvement of microRNAs in epigallocatechin gallate-mediated UVB protection in human dermal fibroblasts. *Oncol Rep* 29(1):253–9.

Alakurtti, S., T. Mäkelä, S. Koskimies, and J. Yli-Kauhaluoma. 2006. Pharmacological properties of the ubiquitous natural product betulin. *Eur J Pharm Sci* 29 (1):1–13.

Alexandrov, P. N., Y. Zhao, B. M. Jones, S. Bhattacharjee, and W. J. Lukiw. 2013. Expression of the phagocytosis-essential protein TREM2 is down-regulated by an aluminum-induced miRNA-34a in a murine microglial cell line. *J Inorg Biochem* 128:267–9.

Alves, R. R., and H. N. Alves. 2011. The faunal drugstore: animal-based remedies used in traditional medicines in Latin America. *J Ethnobiol Ethnomed* 7;7–9.

Bai, T., D. S. Dong, and L. Pei. 2014. Synergistic antitumor activity of resveratrol and miR-200c in human lung cancer. *Oncol Rep* 31 (5):2293–7.

Badal, S., S. A. Williams, G. Huang, S. Francis, P. Vedantam, O. Dunbar, H. Jacobs, T. J. Tzeng, J. Gangemi, R. Delgoda. 2011. Cytochrome P450 1 enzyme inhibition and anticancer potential of chromene amides from Amyris plumieri. *Fitoterapia* 82(2):230–6.

Bartel, D. P. 2004. MicroRNAs: genomics, biogenesis, mechanism, and function. *Cell* 116 (2):281–97.

Bérdy, J. 2005. Bioactive microbial metabolites. *J Antibiot (Tokyo)* 58 (1):1–26.

Bertolini, A., A. Ferrari, A. Ottani, S. Guerzoni, R. Tacchi, and S. Leone. 2006. Paracetamol: new vistas of an old drug. *CNS Drug Rev* 12 (3–4):250–75.

Boesch-Saadatmandi, C., A. Loboda, A. E. Wagner et al. 2011. Effect of quercetin and its metabolites isorhamnetin and quercetin-3-glucuronide on inflammatory gene expression: role of miR-155. *J Nutr Biochem* 22 (3):293–9.

Chakrabarti, M., W. Ai, N. L. Banik, and S. K. Ray. 2013. Overexpression of miR-7-1 increases efficacy of green tea polyphenols for induction of apoptosis in human malignant neuroblastoma SH-SY5Y and SK-N-DZ cells. *Neurochem Res* 38 (2):420–32.

Chakrabarti, M., N. L. Banik, and S. K. Ray. 2013. miR-138 overexpression is more powerful than hTERT knockdown to potentiate apigenin for apoptosis in neuroblastoma in vitro and in vivo. *Exp Cell Res* 319 (10):1575–85.

Chiyomaru, T., S. Yamamura, S. Fukuhara et al. 2013. Genistein up-regulates tumor suppressor microRNA-574-3p in prostate cancer. *PLoS One* 8 (3):e58929.

Chiyomaru, T., S. Yamamura, M. S. Zaman et al. 2012. Genistein suppresses prostate cancer growth through inhibition of oncogenic microRNA-151. *PLoS One* 7 (8):e43812.

Cho, Y. A., W. Lee, and J. S. Choi. 2012. Effects of curcumin on the pharmacokinetics of tamoxifen and its active metabolite, 4-hydroxytamoxifen, in rats: possible role of CYP3A4 and P-glycoprotein inhibition by curcumin. *Pharmazie* 67(2):124–30.

Del Follo-Martinez, A., N. Banerjee, X. Li, S. Safe, and S. Mertens-Talcott. 2013. Resveratrol and quercetin in combination have anticancer activity in colon cancer cells and repress oncogenic microRNA-27a. *Nutr Cancer* 65 (3):494–504.

Dev, S. 1999. Ancient-modern concordance in Ayurvedic plants: some examples. *Environ Health Perspect* 107 (10):783–9.

Feuerhahn, S., C. Giraudon, M. Martínez-Díez et al. 2011. XPF-dependent DNA breaks and RNA polymerase II arrest induced by antitumor DNA interstrand crosslinking-mimetic alkaloids. *Chem Biol* 18 (8):988–99.

Fix, L. N., M. Shah, T. Efferth, M. A. Farwell, and B. Zhang. 2010. MicroRNA expression profile of MCF-7 human breast cancer cells and the effect of green tea polyphenon-60. *Cancer Genomics Proteomics* 7 (5):261–77.

García, M. J., L. P. Saucedo-Cuevas, I. Muñoz-Repeto et al. 2013. Analysis of DNA repair-related genes in breast cancer reveals CUL4A ubiquitin ligase as a novel biomarker of trabectedin response. *Mol Cancer Ther* 12 (4):530–41.

Han, Y. L., H. L. Yu, D. Li et al. 2011. Inhibitory effects of limonin on six human cytochrome P450 enzymes and P-glycoprotein in vitro. *Toxicol In Vitro* 25(8):1828–33.

Heckmann, J. G., M. Dütsch, B. Neundörfer, F. Dütsch, and U. Hartung. 2005. Leech therapy in the treatment of median nerve compression due to forearm haematoma. *J Neurol Neurosurg Psychiatry* 76 (10):1465.

Hegde, V. L., S. Tomar, A. Jackson et al. 2013. Distinct microRNA expression profile and targeted biological pathways in functional myeloid-derived suppressor cells induced by Δ9-tetrahydrocannabinol in vivo: regulation of CCAAT/enhancer-binding protein α by microRNA-690. *J Biol Chem* 288 (52):36810–26.

Hirata, H., K. Ueno, K. Nakajima et al. 2013. Genistein downregulates onco-miR-1260b and inhibits Wnt-signalling in renal cancer cells. *Br J Cancer* 108 (10):2070–8.

Howell, J. C., E. Chun, A. N. Farrell et al. 2013. Global microRNA expression profiling: curcumin (diferuloylmethane) alters oxidative stress-responsive microRNAs in human ARPE-19 cells. *Mol Vis* 19:544–60.

Ji, H. F., X. J. Li, and H. Y. Zhang. 2009. Natural products and drug discovery. Can thousands of years of ancient medical knowledge lead us to new and powerful drug combinations in the fight against cancer and dementia? *EMBO Rep* 10 (3):194–200.

Jiang, R., S. Yamaori, S. Takeda et al. 2011. Identification of cytochrome P450 enzymes responsible for metabolism of cannabidiol by human liver microsomes. *Life Sci* 89(5–6):165–70.

Kang, J., E. Kim, W. Kim et al. 2013. Rhamnetin and cirsiliol induce radiosensitization and inhibition of epithelial-mesenchymal transition (EMT) by miR-34a-mediated suppression of Notch-1 expression in non-small cell lung cancer cell lines. *J Biol Chem* 288 (38):27343–57.

Kaymaz, B. T., V. B. Cetintaş, C. Aktan, and B. Kosova. 2014. MicroRNA-520a-5p displays a therapeutic effect upon chronic myelogenous leukemia cells by targeting STAT3 and enhances the anticarcinogenic role of capsaicin. *Tumour Biol* 35(9):8733–42.

Korish, A. A. 2014. The antidiabetic action of camel milk in experimental type 2 diabetes mellitus: an overview on the changes in incretin hormones, insulin resistance, and inflammatory cytokines. *Horm Metab Res* 46 (6):404–11.

Korish, A. A., and M. M. Arafah. 2013. Camel milk ameliorates steatohepatitis, insulin resistance and lipid peroxidation in experimental non-alcoholic fatty liver disease. *BMC Complement Altern Med* 13:264.

Kronski, E., M. E. Fiori, O. Barbieri et al. 2014. miR181b is induced by the chemopreventive polyphenol curcumin and inhibits breast cancer metastasis via down-regulation of the inflammatory cytokines CXCL1 and -2. *Mol Oncol* 8 (3):581–95.

Kumazaki, M., S. Noguchi, Y. Yasui et al. 2013. Anti-cancer effects of naturally occurring compounds through modulation of signal transduction and miRNA expression in human colon cancer cells. *J Nutr Biochem* 24 (11):1849–58.

Kusuda, R., F. Cadetti, M. I. Ravanelli et al. 2011. Differential expression of microRNAs in mouse pain models. *Mol Pain* 7:17.

Lane, J. A., J. O'Callaghan, S. D. Carrington, and R. M. Hickey. 2013. Transcriptional response of HT-29 intestinal epithelial cells to human and bovine milk oligosaccharides. *Br J Nutr* 110 (12):2127–37.

Leweke, F. M., D. Piomelli, F. Pahlisch et al. 2012. Cannabidiol enhances anandamide signaling and alleviates psychotic symptoms of schizophrenia. *Transl Psychiatry* 2:e94.

Li, X., W. Xie, C. Xie et al. 2014. Curcumin modulates miR-19/PTEN/AKT/p53 axis to suppress bisphenol A-induced MCF-7 breast cancer cell proliferation. *Phytother Res* 28(10):1553–60.

Liu, X., I. Jutooru, P. Lei et al. 2012. Betulinic acid targets YY1 and ErbB2 through cannabinoid receptor-dependent disruption of microRNA-27a:ZBTB10 in breast cancer. *Mol Cancer Ther* 11 (7):1421–31.

Ma, J., L. Cheng, H. Liu et al. 2013. Genistein down-regulates miR-223 expression in pancreatic cancer cells. *Curr Drug Targets* 14 (10):1150–6.

Mahawar, M. M., and D. P. Jaroli. 2007. Traditional knowledge on zootherapeutic uses by the Saharia tribe of Rajasthan, India. *J Ethnobiol Ethnomed* 5;3:25.

Mahomoodally, M. F., and D. P. Sreekeesoon. 2014. A quantitative ethnopharmacological documentation of natural pharmacological agents used by pediatric patients in mauritius. *Biomed Res Int* 2014:136757.

Matsuda, L. A., S. J. Lolait, M. J. Brownstein, A. C. Young, and T. I. Bonner. 1990. Structure of a cannabinoid receptor and functional expression of the cloned cDNA. *Nature* 346 (6284):561–4.

Michalsen, A., S. Klotz, R. Lüdtke, S. Moebus, G. Spahn, and G. J. Dobos. 2003. Effectiveness of leech therapy in osteoarthritis of the knee: a randomized, controlled trial. *Ann Intern Med* 139 (9):724–30.

Molinski, T. F., and B. I. Morinaka. 2012. Integrated approaches to the configurational assignment of marine natural products. *Tetrahedron* 68 (46):9307–43.

Newman, D. J., and G. M. Cragg. 2007. Natural products as sources of new drugs over the last 25 years. *J Nat Prod* 70 (3):461–77.

Newman, D. J., and G. M. Cragg. 2014. Marine-sourced anti-cancer and cancer pain control agents in clinical and late preclinical development. *Mar Drugs* 12 (1):255–78.

Ohno, M., C. Shibata, T. Kishikawa et al. 2013. The flavonoid apigenin improves glucose tolerance through inhibition of microRNA maturation in miRNA103 transgenic mice. *Sci Rep* 3:2553.

Pastrakuljic, A., B. K. Tang, E. A. Roberts, and W. Kalow. 1997. Distinction of CYP1A1 and CYP1A2 activity by selective inhibition using fluvoxamine and isosafrole. *Biochem Pharmacol* 53 (4):531–8.

Pletsch, D., C. Ulrich, M. Angelini, G. Fernandes, and D. S. Lee. 2013. Mothers' "liquid gold": a quality improvement initiative to support early colostrum delivery via oral immune therapy (OIT) to premature and critically ill newborns. *Nurs Leadersh (Tor Ont)* 26:34–42.

Rastogi, S., and P. Chaudhari. 2014. Pigment reduction in nevus of Ota following leech therapy. *J Ayurveda Integr Med* 5 (2):125–8.

Sakano, K., and S. Kawanishi. 2002. Metal-mediated DNA damage induced by curcumin in the presence of human cytochrome P450 isozymes. *Arch Biochem Biophys* 405(2):223–30.

Shankar, S., D. Nall, S. N. Tang et al. 2011. Resveratrol inhibits pancreatic cancer stem cell characteristics in human and KrasG12D transgenic mice by inhibiting pluripotency maintaining factors and epithelial-mesenchymal transition. *PLoS One* 6 (1):e16530.

Shields, M., U. Niazi, S. Badal, T. Yee, M. J. Sutcliffe, and R. Delgoda. 2009. Inhibition of CYP1A1 by quassinoids found in Picrasma excelsa. *Planta Med* 75 (2):137–41.

Shushizadeh, M. R. 2014. Persian gulf bioactive natural drugs. *Jundishapur J Nat Pharm Prod* 9 (2):e19354.

Sinning, C., B. Watzer, O. Coste et al. 2008. New analgesics synthetically derived from the paracetamol metabolite N-(4-hydroxyphenyl)-(5Z,8Z,11Z,14Z)-icosatetra-5,8,11,14-enamide. *J Med Chem* 51 (24):7800–5.

Somasundaram, S., and B. Oommen. 2013. *Antioxidant Flavonoids for Arthritis Treatment: Human and Animal Models*. Edited by R. R. Watson and V. R. Preedy. 1st Edition. London: Elsevier.

Stout, S. M., and N. M. Cimino. 2014. Exogenous cannabinoids as substrates, inhibitors, and inducers of human drug metabolizing enzymes: a systematic review. *Drug Metab Rev* 46(1):86–95

Sun, M., Z. Estrov, Y. Ji, K. R. Coombes, D. H. Harris, and R. Kurzrock. 2008. Curcumin (diferuloylmethane) alters the expression profiles of microRNAs in human pancreatic cancer cells. *Mol Cancer Ther* 7 (3):464–73.

Tsang, W. P., and T. T. Kwok. 2010. Epigallocatechin gallate up-regulation of miR-16 and induction of apoptosis in human cancer cells. *J Nutr Biochem* 21 (2):140–6.

Wang, H., S. Bian, and C. S. Yang. 2011. Green tea polyphenol EGCG suppresses lung cancer cell growth through upregulating miR-210 expression caused by stabilizing HIF-1α. *Carcinogenesis* 32 (12):1881–9.

Whitaker, I. S., O. Oboumarzouk, W. M. Rozen et al. 2012. The efficacy of medicinal leeches in plastic and reconstructive surgery: a systematic review of 277 reported clinical cases. *Microsurgery* 32 (3):240–50.

Xia, J., L. Cheng, C. Mei et al. 2014. Genistein inhibits cell growth and invasion through regulation of MiR-27a in pancreatic cancer cells. *Curr Pharm Des* 20(33):5348–53.

Xia, J., Q. Duan, A. Ahmad et al. 2012. Genistein inhibits cell growth and induces apoptosis through up-regulation of miR-34a in pancreatic cancer cells. *Curr Drug Targets* 13 (14):1750–6.

Xu, L., J. Xiang, J. Shen et al. 2013. Oncogenic MicroRNA-27a is a target for genistein in ovarian cancer cells. *Anticancer Agents Med Chem* 13 (7):1126–32.

Yang, C. H., J. Yue, M. Sims, and L. M. Pfeffer. 2013. The curcumin analog EF24 targets NF-κB and miRNA-21, and has potent anticancer activity in vitro and in vivo. *PLoS One* 8 (8):e71130.

Zaman, M. S., S. Thamminana, V. Shahryari et al. 2012. Inhibition of PTEN gene expression by oncogenic miR-23b-3p in renal cancer. *PLoS One* 7 (11):e50203.

Zhai, S., R. Dai, F. K. Friedman, and R. E. Vestal. 1998. Comparative inhibition of human cytochromes P450 1A1 and 1A2 by flavonoids. *Drug Metab Dispos* 26 (10):989–92.

Zhao, Y. N., W. F. Li, F. Li et al. 2013. Resveratrol improves learning and memory in normally aged mice through microRNA-CREB pathway. *Biochem Biophys Res Commun* 435 (4):597–602.

Zheng, J., C. Wu, Z. Lin et al. 2014. Curcumin up-regulates phosphatase and tensin homologue deleted on chromosome 10 through microRNA-mediated control of DNA methylation–a novel mechanism suppressing liver fibrosis. *FEBS J* 281 (1):88–103.

Zhong, Z., Z. Dong, L. Yang, X. Chen, and Z. Gong. 2012. Inhibition of proliferation of human lung cancer cells by green tea catechins is mediated by upregulation of let-7. *Exp Ther Med* 4 (2):267–72.

Appendix I

Supplementary Data Set 1: Genes of Interest to Be Studied

Symbol	Description	Gene Name	RT2 Catalog
APC	Adenomatous polyposis coli	BTPS2/DP2/DP2.5/DP3/GS	PPH00070B
BRMS1	Breast cancer metastasis suppressor 1	DKFZp564A063	PPH01052E
CCL7	Chemokine (C-C motif) ligand 7	FIC/MARC/MCP-3/MCP3/ MGC138463/MGC138465/NC28/ SCYA6/SCYA7	PPH00575B
CD44	CD44 molecule (Indian blood group)	CDW44/CSPG8/ECMR-III/HCELL/ IN/LHR/MC56/MDU2/MDU3/ MGC10468/MIC4/MUTCH-I/Pgp1	PPH00114A
CDH1	Cadherin 1, type 1, E-cadherin (epithelial)	Arc-1/CD324/CDHE/ECAD/LCAM/ UVO	PPH00135E
CDH11	Cadherin 11, type 2, OB-cadherin (osteoblast)	CAD11/CDHOB/OB/OSF-4	PPH00667A
CDH6	Cadherin 6, type 2, K-cadherin (fetal kidney)	KCAD	PPH00657B
CDKN2A	Cyclin-dependent kinase inhibitor 2A (melanoma, p16, inhibits CDK4)	ARF/CDK4I/CDKN2/CMM2/INK4/ INK4a/MLM/MTS1/TP16/p14/ p14ARF/p16/p16INK4/p16INK4a/p19	PPH00207B
CHD4	Chromodomain helicase DNA-binding protein 4	DKFZp686E06161/Mi-2b/Mi2-BETA	PPH02273A
COL4A2	Collagen, type IV, alpha 2	DKFZp686I14213/FLJ22259	PPH00247E
CST7	Cystatin F (leukocystatin)	CMAP	PPH05560E
CTBP1	C-terminal-binding protein 1	BARS/MGC104684	PPH02774A
CTNNA1	Catenin (cadherin-associated protein), alpha 1, 102kDa	CAP102/FLJ36832/FLJ52416	PPH00646B
CTSK	Cathepsin K	CTS02/CTSO/CTSO1/CTSO2/ MGC23107/PKND/PYCD	PPH02137A
CTSL1	Cathepsin L1	CATL/CTSL/FLJ31037/MEP	PPH00113E
CXCL12	Chemokine (C-X-C motif) ligand 12 (stromal cell-derived factor 1)	PBSF/SCYB12/SDF-1a/SDF-1b/ SDF1/SDF1A/SDF1B/TLSF-a/ TLSF-b/TPAR1	PPH00528B
CXCR4	Chemokine (C-X-C motif) receptor 4	CD184/D2S201E/FB22/HM89/ HSY3RR/LAP3/LCR1/LESTR/ NPY3R/NPYR/NPYRL/NPYY3R/ WHIM	PPH00621A
DENR	Density-regulated protein	DRP/DRP1/SMAP-3	PPH08394B
EPHB2	EPH receptor B2	CAPB/DRT/EPHT3/ERK/Hek5/ MGC87492/PCBC/Tyro5	PPH05673E

(Continued)

199

Supplementary Data Set 1: Genes of Interest to Be Studied (*Continued*)

Symbol	Description	Gene Name	RT2 Catalog
ETV4	Ets variant 4	E1A-F/E1AF/PEA3/PEAS3	PPH01099E
EWSR1	Ewing sarcoma breakpoint region 1	EWS	PPH18254E
FAT1	FAT tumor suppressor homolog 1 (Drosophila)	CDHF7/FAT/ME5/hFat1	PPH00403A
FGFR4	Fibroblast growth factor receptor 4	CD334/JTK2/MGC20292/TKF	PPH00390A
FLT4	Fms-related tyrosine kinase 4	FLT41/LMPH1A/PCL/VEGFR3	PPH00803E
FN1	Fibronectin 1	CIG/DKFZp686F10164/ DKFZp686H0342/DKFZp686I1370/ DKFZp686O13149/ED-B/FINC/FN/ FNZ/GFND/GFND2/LETS/MSF	PPH00143B
FXYD5	FXYD domain containing ion transport regulator 5	HSPC113/IWU-1/IWU1/KCT1/OIT2/ PRO6241/RIC/dysad	PPH17099E
GNRH1	Gonadotropin-releasing hormone 1 (luteinizing-releasing hormone)	GNRH/GRH/LHRH/LNRH	PPH02233A
KISS1R	KISS1 receptor	AXOR12/GPR54/HOT7T175	PPH14411A
HGF	Hepatocyte growth factor (hepapoietin A; scatter factor)	F-TCF/HGFB/HPTA/SF	PPH00163B
HPSE	Heparanase	HPA/HPR1/HPSE1/HSE1	PPH01060E
HRAS	V-Ha-ras Harvey rat sarcoma viral oncogene homolog	C-BAS/HAS/C-H-RAS/C-HA-RAS1/ CTLO/H-RASIDX/HAMSV/ HRAS1/K-RAS/N-RAS/RASH1	PPH00159B
HTATIP2	HIV-1 Tat interactive protein 2, 30 kDa	CC3/FLJ26963/SDR44U1/TIP30	PPH06957A
IGF1	Insulin-like growth factor 1 (somatomedin C)	IGF1A/IGFI	PPH00167B
IL18	Interleukin 18 (interferon-gamma-inducing factor)	IGIF/IL-18/IL-1g/IL1F4/MGC12320	PPH00580B
IL1B	Interleukin 1, beta	IL-1/IL1-BETA/IL1F2	PPH00171B
IL8RB	Interleukin 8 receptor, beta	CD182/CDw128b/CMKAR2/CXCR2/ IL8R2/IL8RA	PPH00608E
ITGA7	Integrin, alpha 7	FLJ25220	PPH00665A
ITGB3	Integrin, beta 3 (platelet glycoprotein IIIa, antigen CD61)	CD61/GP3A/GPIIIa	PPH00178C
CD82	CD82 molecule	4F9/C33/GR15/IA4/KAI1/R2/SAR2/ ST6/TSPAN27	PPH01312A
KISS1	KiSS-1 metastasis-suppressor	KiSS-1/METASTIN/MGC39258	PPH01080A
KRAS	V-Ki-ras2 Kirsten rat sarcoma viral oncogene homolog	C-K-RAS/K-RAS2A/K-RAS2B/K-RAS4A/K-RAS4B/KI-RAS/KRAS1/ KRAS2/NS3/RASK2	PPH00181E

(Continued)

Supplementary Data Set 1: Genes of Interest to Be Studied (*Continued*)

Symbol	Description	Gene Name	RT2 Catalog
RPSA	Ribosomal protein SA	37LRP/67LR/LAMBR/LAMR1/LRP/ p40	PPH18264A
MCAM	Melanoma cell adhesion molecule	CD146/MUC18	PPH00651A
MDM2	Mdm2 p53-binding protein homolog (mouse)	HDMX/MGC5370/MGC71221/hdm2	PPH00193E
MET	Met proto-oncogene (hepatocyte growth factor receptor)	AUTS9/HGFR/RCCP2/c-Met	PPH00194A
METAP2	Methionyl aminopeptidase 2	MAP2/MNPEP/p67/p67eIF2	PPH12531E
MGAT5	Mannosyl (alpha-1,6-)- glycoprotein beta-1,6-*N*-acetyl- glucosaminyltransferase	GNT-V/GNT-VA	PPH01059E
MMP10	Matrix metallopeptidase 10 (stromelysin 2)	SL-2/STMY2	PPH00896B
MMP11	Matrix metallopeptidase 11 (stromelysin 3)	SL-3/ST3/STMY3	PPH00236B
MMP13	Matrix metallopeptidase 13 (collagenase 3)	CLG3	PPH00121B
MMP2	Matrix metallopeptidase 2 (gelatinase A, 72 kDa gelatinase, 72 kDa type IV collagenase)	CLG4/CLG4A/MMP-II/MONA/ TBE-1	PPH00151B
MMP3	Matrix metallopeptidase 3 (stromelysin 1, progelatinase)	CHDS6/MGC126102/MGC126103/ MGC126104/MMP-3/SL-1/STMY/ STMY1/STR1	PPH00235E
MMP7	Matrix metallopeptidase 7 (matrilysin, uterine)	MMP-7/MPSL1/PUMP-1	PPH00809E
MMP9	Matrix metallopeptidase 9 (gelatinase B, 92 kDa gelatinase, 92 kDa type IV collagenase)	CLG4B/GELB/MMP-9	PPH00152E
MTA1	Metastasis associated 1	–	PPH01083E
MTSS1	Metastasis suppressor 1	DKFZp781P2223/FLJ44694/ KIAA0429/MIM/MIMA/MIMB	PPH10073A
MYC	V-myc myelocytomatosis viral oncogene homolog (avian)	MRTL/bHLHe39/c-Myc	PPH00100A
MYCL1	V-myc myelocytomatosis viral oncogene homolog 1, lung carcinoma derived (avian)	LMYC/MYCL/bHLHe38	PPH00182B
NF2	Neurofibromin 2 (merlin)	ACN/BANF/SCH	PPH00203A
NME1	Non-metastatic cells 1, protein (NM23A) expressed in	AWD/GAAD/NB/NBS/NDPK-A/ NDPKA/NM23/NM23-H1	PPH01314A

(Continued)

Supplementary Data Set 1: Genes of Interest to Be Studied (*Continued*)

Symbol	Description	Gene Name	RT2 Catalog
NME2	Non-metastatic cells 2, protein (NM23B) expressed in	MGC111212/NDPK-B/NDPKB/ NM23-H2/NM23B/puf	PPH20180B
NME4	Non-metastatic cells 4, protein expressed in	NDPK-D/NM23H4/nm23-H4	PPH01086A
NR4A3	Nuclear receptor subfamily 4, group A, member 3	CHN/CSMF/MINOR/NOR1/TEC	PPH00411A
PLAUR	Plasminogen activator, urokinase receptor	CD87/UPAR/URKR	PPH00797B
PNN	Pinin, desmosome associated protein	DRS/SDK3/memA/pinin	PPH19485E
PTEN	Phosphatase and tensin homolog	10q23del/BZS/MGC11227/MHAM/ MMAC1/PTEN1/TEP1	PPH00327E
RB1	Retinoblastoma 1	OSRC/RB/p105-Rb/pRb/pp110	PPH00228E
RORB	RAR-related orphan receptor B	NR1F2/ROR-BETA/RZR-BETA/ RZRB/bA133M9.1	PPH05876E
SET	SET nuclear oncogene	2PP2A/I2PP2A/IGAAD/IPP2A2/ PHAPII/TAF-I/TAF-IBETA	PPH20624E
SMAD2	SMAD family member 2	JV18/JV18-1/MADH2/MADR2/ MGC22139/MGC34440/hMAD-2/ hSMAD2	PPH01949E
SMAD4	SMAD family member 4	DPC4/JIP/MADH4	PPH00134B
SRC	V-src sarcoma (Schmidt-Ruppin A-2) viral oncogene homolog (avian)	ASV/SRC1/c-SRC/p60-Src	PPH00103C
SSTR2	Somatostatin receptor 2	–	PPH02064A
SYK	Spleen tyrosine kinase	DKFZp313N1010/FLJ25043/ FLJ37489	PPH01639E
TCF20	Transcription factor 20 (AR1)	AR1/KIAA0292/SPBP	PPH14308E
TGFB1	Transforming growth factor, beta 1	CED/DPD1/TGFB/TGFbeta	PPH00508A
TIMP2	TIMP metallopeptidase inhibitor 2	CSC-21K	PPH00904A
TIMP3	TIMP metallopeptidase inhibitor 3	HSMRK222/K222/K222TA2/SFD	PPH00762A
TIMP4	TIMP metallopeptidase inhibitor 4	–	PPH00889E
TNFSF10	Tumor necrosis factor (ligand) superfamily, member 10	APO2L/Apo-2L/CD253/TL2/TRAIL	PPH00242E
TP53	Tumor protein p53	FLJ92943/LFS1/TRP53/p53	PPH00213E
TRPM1	Transient receptor potential cation channel, subfamily M, member 1	LTRPC1/MLSN1	PPH12993A
TSHR	Thyroid stimulating hormone receptor	CHNG1/LGR3/MGC75129/hTSHR-I	PPH02344B

(*Continued*)

Supplementary Data Set 1: Genes of Interest to Be Studied (*Continued*)

Symbol	Description	Gene Name	RT2 Catalog
VEGFA	Vascular endothelial growth factor A	MGC70609/MVCD1/VEGF/ VEGF-A/VPF	PPH00251B
B2M	Beta-2-microglobulin	–	PPH01094E
HPRT1	Hypoxanthine phosphoribosyltransferase 1	HGPRT/HPRT	PPH01018B
RPL13A	Ribosomal protein L13a	–	PPH01020B
GAPDH	Glyceraldehyde-3-phosphate dehydrogenase	G3PD/GAPD/MGC88685	PPH00150E
ACTB	Actin, beta	PS1TP5BP1	PPH00073E
HGDC	Human Genomic DNA Contamination	HIGX1A	
RTC	Reverse transcription control	RTC	
RTC	Reverse transcription control	RTC	
RTC	Reverse transcription control	RTC	
PPC	Positive PCR Control	PPC	
PPC	Positive PCR Control	PPC	
PPC	Positive PCR Control	PPC	

Index